T0251335

DAMS AND DIKES IN DEVELOPMENT

The problem, though, is not the dams.
It is the hunger.
It is the thirst.
It is the darkness of a township.
It is township and rural huts without running water, lights or sanitation.
It is the time wasted in gathering water by hand.
There is a real pressing need for power in every sense of the word.

Nelson Mandela, 16 November 2000, London

DAMS AND DIKES IN DEVELOPMENT

PROCEEDINGS OF THE SYMPOSIUM AT THE OCCASION OF THE WORLD
WATER DAY, 22 MARCH 2001

EDITED BY:

Hans van Duivendijk, Bart Schultz and Cees-Jan van Westen

CRC Press
Taylor & Francis Group
Boca Raton London New York

CRC Press is an imprint of the
Taylor & Francis Group, an **informa** business

Library of Congress Cataloging-in-Publication Data

Applied for

Printed in the Netherlands by Grafisch Produktiebedrijf Gorter, Steenwijk

Published by: A.A. Balkema Publishers, a member of Swets & Zeitlinger Publishers
www.balkema.nl and www.szp.swets.nl

ISBN 90 5809 541 X

Preface

In the light of the global debate on dams, and the national debate on the future policy on flood protection and water management, the Netherlands National Committee on Large Dams (NETHCOLD) and the Netherlands Committee of the International Commission on Irrigation and Drainage (NETHCID) jointly organised a one-day symposium on the theme 'Dams and Dikes in Development'. The symposium was organised in co-operation with:

- the Royal Institution of Engineers in The Netherlands (KIVI) and more in particular its divisions 'Bouw- en Waterbouwkunde' and 'Water Management'
- the 'Dispuut Watermanagement' and the 'Waterbouw Dispuut' of the Faculty of Civil Engineering, Delft University of Technology

During this symposium the role of dikes and dams in the management and development of river basins was presented and discussed. Attention was paid to their role in The Netherlands, as well as internationally.

NETHCID promotes that annually at the occasion of the World Water Day - 22 March - an event is organised related to one of the important issues regarding water management. This is the sixth event in this series.

NETHCOLD, as the national committee of the International Commission on Large Dams (ICOLD), has as its aim the promotion of Dutch know how and expertise, developed during the last century on integrated river basin and delta development, in the international field of large dams. In the past NETHCOLD has also, from time to time, organised symposia at the Delft University of Technology on typical large dams issues which are of interest to future civil, hydraulic and environmental engineers.

We hope that these proceedings will be of interest for you.

Delft, 15 September 2002

Hans van Duivendijk
Chairman NETHCOLD

Bart Schultz
Netherlands representative ICID

Preface

Contents

Introduction

IR. HANS VAN DUIVENDIJK

Ladies and Gentlemen,

A very warm welcome to you all. A welcome on behalf of the Netherlands Committees on Irrigation and Drainage and that on Large Dams respectively.

During the past years NETHID, that is the Netherlands Committee of ICID (which stands for International Commission on Irrigation and Drainage) has always organised symposia at the occasion of the annual World Water Day while NETHCOLD, that is the Netherlands Committee of ICOLD (which in turn stands for International Commission on Large Dams) has organised a number of symposia at Delft University.

More recently, these two international organisations in the field of water and its utilisation and application for the benefit of mankind, work, as far as dams are concerned, more closely together and that, on its own, was already a good reason for holding this Symposium. Moreover, in November last year, an important report on dams was launched by another, temporary, organisation called the 'World Commission on Dams' (WCD). I am sure, that by now, you are completely confused about all these commissions and committees.

Today, you will hear from our various eminent speakers what are the concerns and benefits of dams and, hopefully, by the end of the day you have an idea why dams, and also dikes, are needed, why, in a number of cases and places, people are against dams and, last but not least, how such controversial situations should be solved.

You have already heard that we will not only speak about dams but also about dikes. Consequently, the symposium was first called 'Dams and Dikes in Development' which prompted a good friend of mine, Mr. van den Berg, a former secretary of NETHCOLD, to say: 'typically Dutch to mention dikes before dams, can't you people think internationally?'

Well, I hope that today we are able to demonstrate to you that we do think internationally and, moreover, that we are not only interested to promote our Dutch *technical* know how in the field of dams and dikes but even more our approach and experience in non- technical issues concerning these important hydraulic engineering structures.

The fact that we are guest today of the International Institute for Infrastructural, Hydraulic and Environmental Engineering (IHE), shows already that we are prepared to look at these structures from various angles in an international context.

Dikes and dams have similar functions in the field of water: they defend us against floods and storm surges, they enable us to use water for human consumption and for food production and both may be needed when we use stored water for hydro power.

1

Dams and dikes – benefits, costs and option assessment

PROF. KAARE HOEGE

The program prepared for this symposium on the future policy on flood protection and water resource management promises to give us a most interesting and educational day. I will start out with an overview and a few statistics on dams and dikes, and a discussion of some of the important aspects of benefits and costs, risk analysis and decision making under uncertainty. Then I will reflect on the outcome of the important work of the 2-year World Commission on Dams (WCD), which recently completed its terms of reference and issued its final report.

Hopefully, my remarks will serve as a useful framework for the many interesting lectures and discussion periods to follow.

Background and some statistics

- the world population is increasing - in some regions dramatically. It is estimated that by year 2025, the population will have increased from the present 6 billions to 8 billions, and much of the increase will take place in water-scarce regions in developing countries;
- freshwater resources are limited and unevenly distributed, and seasonal variations and climatic irregularities seriously impede the efficient use of river runoff;
- drought on the one hand and river flooding on the other occur with catastrophic consequences which the world community is struggling to mitigate;
- the rise of the sea level combined with land subsidence and ocean storms cause flooding of near-shore areas that require protective dikes in many parts of the world.

In an effort to alleviate some of the 'shortcomings' and threats of nature, and also to provide energy and electricity, man has built an essential infrastructure consisting of dams and dikes (in the following simply referred to as dams):

- 45,000 large dams (defined as dams higher than 15 m or 5 - 15 m high with reservoirs larger than 3 million m^3). About 23,000 of these are located in China;
- an estimated 800,000 smaller dams;
- 70% of all large dams are less than 30 m high;
- less than 1% of all large dams are over 100 m high.

I mention these last two points because reports in the media and discussions in different fora sometimes give the impression that most large dams are 50 - 100 m high or higher.

The primary functions of existing large dams are:

– Irrigation for agricultural purposes, mainly for food supply	48%
– Electricity generation	20%
– Water supply	15%
– Flood protection	8%
– Recreation and improvement of the environment	4%
– Inland navigation	2%
– Fish breeding	2%
– Fire fighting and miscellaneous	1%

Very many of the dams are multi-purpose, serving more than one of the functions listed above. This is an essential aspect, and careful planning makes this characteristic more and more common in recent dams.

Costs and social inequity

The construction and operation of any infrastructure for the benefit of the population and a country at large, have indirect costs and disadvantages associated with them. The penalties are primarily felt locally, and this creates an imbalance. The issues of social equity and the need for improvements in local living conditions beyond mere compensation in material goods, are essential. Careful planning and follow-up are required by the decision makers, including consultations with and involvement of all stakeholders, option assessment and risk evaluation.

In the case of dike, dam and reservoir building, it is often required that local people be resettled, and there may be significant socio-economic and environmental impacts both upstream and downstream of the dam. The International Commission on Large Dams (ICOLD) has since 1968 a standing committee to study the effects of dams on the environment and provide guidance on the mitigation of any negative effects. Several ICOLD Congress Sessions and Special Symposia have been organised to focus on these essential issues in addition to safety and economy. In 1995 ICOLD issued its *Position Paper on Dams and Environment* (second printing in 1997) to summarise and underline its policy.

Needs, options and decision making

When the critical needs are water supply for food production, drinking and sanitary purposes, flood protection and electricity, what realistic options does society have? What are the short and long-term consequences of the different options?

The 'do-nothing' alternative (inaction) is one of the available options, and a consequence analysis should be performed for that alternative as well. That often seems to be forgotten. In too many cases one ends up with this option, not because it is decided to represent the best alternative, but due to inability, or fear of criticism, to choose a better, but possibly controversial, alternative.

The director-general of UNESCO, F. Mayor, said at the occasion of the first World Water Forum in 1997: 'New water projects, conceived in an ecologically sensitive way, take 15 - 20 years to plan and implement. If we delay, the pressure to quench the thirst of poor people may force the recourse to ecologically or otherwise unsound projects'.

The many concerns must not lead to paralysis in the conscious decision making. In the long run, that may be the worst outcome for the people affected. No alternative is ideal for all. The decision makers must make sure that there is equity in the distribution of benefits - locally and regionally. Any infrastructure should be planned and engineered to best reflect the needs and values of the diverse societies they are meant to serve.

Probabilistic optimisation and risk analysis

Because there are uncertainties involved, the optimisation process is a probabilistic one. The benefits and direct costs of the different options may be estimated fairly reliably, although climatic changes create operational uncertainties, and so do local geological conditions for the construction of dams and dikes. On the other hand, long-term negative effects on the social and physical environment are uncertain and difficult to quantify. Unfortunately, these latter uncertainties will often dominate the outcome of the comparative analyses among the options.

In any decision there is risk involved. There is also a large difference between a person's acceptance level for voluntary and imposed risks. Studies show that there may be a factor of at least 10^3 between them. At this stage in the decision process, objective evaluation in the debate between pros and cons tends to be given less emphasis and unfounded generalisations, entrenched attitudes and rhetoric tend to take over. Generalisations, often based on only a few specific cases, are particularly harmful, but often effective, in the option assessment process, because each situation is case-specific. There is no such thing as a 'typical' dam or reservoir built under typical upstream and downstream conditions.

We definitely need to develop a more systematic assessment procedure to compare the total quality of the different options proposed - including the status quo option. After the procedure has been agreed upon, one must then stick to it and abide by the outcome.

Some facts seem indisputable

The environmentalist has the underlying philosophy that the natural environment and the associated natural processes, untouched by man, represent the ideal state, and that any change to this state is 'negative'. However, due to population growth, also a natural process, and several other reasons mentioned earlier in this presentation, a situation has evolved where some human-induced changes are required to prevent further human suffering. Incidentally, natural geological processes lead each year to numerous landslides and earthquakes that cause tragedies and dramatic changes of the environment. One of these is changing the course of and sometimes blocking rivers, thus creating inundation upstream and a threat to the people living downstream, because these dams are not properly 'engineered'.

Some indisputable facts seem to be:
- irrigation is no longer only an option, but a necessity for feeding the rapidly increasing population. The efficiency of irrigation must be improved, as harmful

waterlogging and salinity seem to affect approximately 20% of the irrigated land;
- at present, approximately 2 billion people are without reliable access to clean water, a basic human right, and are without electricity;
- groundwater resources are in many areas already overtaxed or polluted. Dams are good instruments to manage surface water resources and to help recharge groundwater aquifers in a sustainable manner;
- solar power, and also wind, wave, tidal and geothermal power represent promising renewable sources of clean energy. In the foreseeable future, however, these options can only, except in a few local regions, provide a small fraction of the levels of energy output required;
- in the meantime, non-renewable and air polluting fossil-fuelled thermal power plants are being built. It is an unfortunate fact that this takes place by importing coal and oil even into those developing countries where hydropower is the only natural energy resource;
- one reason for this trend is that any negative social and environmental effects caused by dams are felt locally, and the costs are included, while the harmful effects of gases and air pollution produced by thermal power are not yet included (internalised) when options are assessed and compared;
- building of dams and reservoirs requires in many cases resettlement of people. However, one cannot generalise. Norway, for instance, has built 330 large dams, and only approximately 20 persons have had to resettle. Some other countries show similar statistics;
- of the 45,000 large dams built, some have certainly not given the anticipated benefits, or the environmental and social costs have been higher than expected. In the light of 'post-factum' wisdom, some of these dams should not have been built.

WCD mandate and objective

The 2-year World Commission on Dams (WCD) was established in May 1998 by the World Bank and the World ConservAation Union (IUCN). The specific terms of reference were:
- review the development effectiveness of large dams and assess alternatives for water resources and energy development;
- develop internationally acceptable criteria, guidelines and standards, where appropriate, for the planning, design, appraisal, construction, operation, monitoring and decommissioning of dams.

During the work of the Commission Chairman Kader Asmal, who was then South Africa's Minister of Water Affairs and Forestry, stated: 'Dams per se are not the problematic issue, but rather the flawed process of decision making that has been at the centre of conflicts associated with some dams'. In most cases the 'flawed process' is due to failure of governance, implementation or follow-up vis-à-vis the local people affected.
    In the preface to the WCD Final Report (November, 2000), Prof. Kader Asmal writes: 'We are much more than a *Dams Commission*. We are a Commission to heal the deep and self-inflicted wounds torn open wherever and whenever far too few determine for far too many how best to develop or use our resources'.
Thus, the scope of work was expanded as the work progressed.

WCD conclusions

In the preface to the WCD Final Report, the Chairman concludes that:

> 'In the following pages we do not endorse globalisation as led from above by a few men. We do endorse globalisation as led from below by all, a new approach to global policy and development'.

After two years of intense work the Commission concludes that:

> 'We have found that unprecedented expansion in large dam building over the past century, harnessing water for irrigation, domestic and industrial consumption, electricity generation and flood control has clearly benefited many people globally. Nonetheless, this positive contribution of large dams to development has been marred in many cases by significant environmental and social impacts which, when viewed from today's values, are unacceptable'.

The Commission also concludes that, in general, development should be based on five objectives, or core values:
- 'Equity in resource allocation and in the spread of benefits;
- Sustainability in the use of the world's diminishing resource base;
- Openness and participation in decision making processes;
- Efficiency in the management of existing infrastructure development;
- Accountability towards present and future generations'.

Any policy on large infrastructure projects - whether for dams, highways, power stations, or other mega-installations - has to be developed in this context'.

Personal reflections on the WCD Final Report

I am very impressed by the volume of work accomplished by the WCD, not the least by the way the work was organised and reported almost within the original time period allotted (2 years). The 'knowledge base' that was created, will be of great value in our future efforts to plan and use dams in water management, flood protection and regional development.

The final report does not spend much time reviewing the many benefits of dams and reservoirs. That is unfortunate. The Commission and its staff must have assumed that the reader is familiar with all those. Allow me, by a similar reasoning, not to spend much time here on all the valuable aspects of the report and the advice and guidance it gives, but rather reflect on some shortcomings and, as I see it, lost opportunities.

Scope of work

The Commission generalised its scope of work by declaring that it was much more than a *Dams Commission*, rather a Commission 'to heal the deep and self-inflicted wounds torn open wherever and whenever far too few determine for far too many how best to develop or use our resources'. The Commission decided to use its unique opportunity to crusade for human rights, in general. This is also the strongest aspect of the report, which hopefully will give positive effects in many areas. However, this allowed less time and resources to be put on the Commission's specific mandate related to the assessment of alternatives to dams and dikes to meet present and future critical needs.

Option assessment

The mandate called for 'assess alternatives for water resources and energy development'. This is an essential task, because as I suggested before, it is the weakest link in our decision process related to water resource and energy developments. Precious little substantial guidance is given in the WCD final report with respect to discussion of alternative systems for water supply (e.g. the use of ground water, decentralised storage, rainfed systems, etc.) and for power production (e.g. coal, oil, gas, nuclear, solar, etc.). It would certainly have been a formidable, difficult and controversial task to tackle. Yes, but it was part of the mandate and a unique funding situation was made available through extensive goodwill towards a Commission given a unique opportunity.

The final report says that there are many alternatives, but fails to elaborate, except presenting general statements like:
– 'use existing supply frugally;
– increase efficiency of existing facilities;
– use demand-side management'.

In affluent regions of the world, we can certainly save on our use of water and energy resources by changing our attitudes, ways and means. And we should, rather than first think of building new facilities. However, that advice is of little help to the developing regions where there is a basic shortage of both these resources, at least for parts of the year, to satisfy critical human and societal needs.

When the WCD report was released in London 16 November 2000, Nelson Mandela was invited to give the keynote speech. He said:

> 'Political freedom alone is still not enough if you lack clean water. Freedom alone is not enough without light to read at night, without access to water to irrigate your farm, without the ability to catch fish to feed your family.'

> 'The problem, though, is not the dams. It is the hunger. It is the thirst. It is the darkness of a township. It is township and rural huts without running water, lights or sanitation. It is the time wasted in gathering water by hand. There is a real pressing need for power in every sense of the word.'

> 'Rather than single out dams for excessive blame or credit, all of us must wrestle with the difficult questions we face. It is one thing to find fault with an existing system. It is another thing altogether, a more difficult task, to replace it with an approach that is better.'

Core values and sense of urgency

The five objectives (core values) identified by the Commission and quoted above, are very well defined and formulated. I do not think anyone will disagree with them. However, when the Commission compares dams with other types of infrastructure, I miss a statement that most dams are built to satisfy critical needs for survival - water. Business as usual is no option. Freshwater scarcity is ranked second only to global warming among the major challenges for the 21st century.

Review of old dam developments

The Commission concludes that 'many dams have had significant environmental and social impacts which, when viewed from today's values, are unacceptable'. That is undoubtedly true in many cases. Our priorities and values have changed significantly since World War II, and I think this very general conclusion could be drawn about much of the infrastructure built 50 years ago. I wish the WCD also had reviewed and assessed dams built during the last 20 years. Thus, some useful conclusions and guidelines could have been given about present planning and implementation practice. That would have been useful and informative. The Commission would have found that in present practice very sincere efforts are being made to mitigate any negative social and environmental effects.

WCD guidelines

WCD presents 5 core values, 7 strategies and 26 guidelines. The guidelines do not seem well focussed and structured compared to the rest of the report, which is exemplary. Furthermore, the development process seems strongly idealised, disregarding many aspects of regional and cultural differences. That is unfortunate, because as stated by Briscoe of the World Bank (March 2001): 'if they (the guidelines) are taken as a check list of requirements to be complied with and conformed to, then they are strongly opposed by all the governments we (i.e. the World Bank) have consulted ....'

The potentially very positive impact of the Commission's work will be severely limited if the countries who will be building most of the dams in the future, do not take the report seriously due to the way the guidelines ('requirements') are put forward.

'Unreasonable' statements

In some instances the authors of parts of the report seem to turn an argument which basically is in favour of dam or dike building into a disadvantage, e.g. 'Dams built for flood protection have led to greater vulnerability to flood hazards due to increased settlement in neighbouring areas still at risk from floods.' If this argument is followed for safety-promotion work in society in general, it may lead to some very strange conclusions and consequences. For the case in hand, the probability of flooding has been reduced by orders of magnitude by the building of a dam. The fact that more people now may move into the neighbourhood because they like it, cannot be an argument against dam building to protect people already living there. The alternative is resettlement of the local people.

Another example of unreasonable wording is the argumentation around greenhouse gases. For some special conditions it has been found that the production of greenhouse gases from reservoir building can be significant, but, in general, hydropower is much superior to the fossil fuel alternative. However, the report presentation leaves the reader bewildered and the media confused.

The forces of globalisation of the economy and the emergence of the free energy market are recognised as important. The tone and wording in parts of the report may be used as an argument for investors to shy away from hydropower and rather invest in environmentally much less favourable energy projects, e.g. coal power.

Final observations

The commissioners conclude by saying:

'We hope that one of the lasting results of the WCD process will have been to change the tenor of the debate about dams from one of lack of trust and destructive confrontation to co-operation, shared goals and more equitable development outcomes'.

This hope is certainly shared by ICOLD who will continue its work using the knowledge base created by WCD. Increased awareness on 'both sides' and a better mutual understanding may already have been achieved through the extensive work of the Commission.

However, the mission of the Commission would have been even more successful if the authors of the report had been more even-handed in their presentations. In this respect the tone of the Executive Summary and of the specially edited Overview are more balanced than the chapters of the main body, which I assume the commissioners had less time to evaluate. At times during the reading, I got the impression that the writer states he is for a fair and systematic option assessment, but really means that dam building is only an option when everything else fails.

After the report release, having met some critical review comments, spokesmen for the Commission respond that we have to see the overall value of the report, grasp the opportunity to plan the way forward, and work together for a common cause. I fully agree, but the Commission should have wrestled with the issue of realistic option assessment, including the status quo option, when it comes to future water resource development. That would have brought us another step forward and made the dam debate more realistic and constructive. As Nelson Mandela said: 'it is one thing to find fault with an existing system. It is another thing altogether, a more difficult task, to replace it with an approach that is better'.

Looking ahead

People will continue to build dams for many purposes, especially in the developing world. The reason is that dams in very many cases will be judged to represent the best option among feasible alternatives to satisfy critical human and societal needs.

A more systematic and rational procedure of option assessment must be developed and practised to facilitate the decision making. Decisions made under uncertainty call for probabilistic evaluations, and the consequences of the 'do-nothing' option must also be critically assessed and compared with the other alternatives.

As the population continues to increase in water-scarce regions, additional water resources must be developed. The only practical way to achieve this on the scale required is to somehow increase storage capacity, combined with methods for reducing waste and pollution and techniques for recycling water.

With our increased knowledge and understanding, dams and dikes will be made socially equitable and environmentally sustainable through stakeholder involvement, monitoring and continuous evaluation.

ICOLD's mission is to continuously improve the state of the art - and we will.

2

Dikes, dams and water management

Prof. dr. Bart Schultz

'The pace of change in our world is speeding up, accelerating to the point
where it threatens to overwhelm the management capacity of political
leaders. This acceleration in history comes not only from advancing
technology, but also from unprecedented world population growth, even
faster economic growth, and the increasingly frequent collisions between
expanding human demands and the limits of the earth's natural systems.'

Lester R. Brown, 1996

Introduction

Dikes, dams and water management are in a direct relation to each other, as well as
with the needs of societies and acceptable side effects. In relation to water
management dikes and dams may have a function for irrigation, drainage, flood
management and flood protection. Dikes have a function for flood protection only,
but their development generally implies the installation of a drainage system to drain
the surplus water from the endiked lands. Two developments are of crucial
importance for the future of dikes, dams and water management:
– based on projections of population growth it is expected that food production
   has to be doubled in the next 25 years. This can only be achieved by significant
   improvements in irrigation and drainage in conjunction with an increase in
   storages (van Hofwegen and Svendsen, 2000);
– it is expected that within 50 years 80% of the world's population will live in
   coastal and deltaic areas. By far the majority of them will live in urban areas.
   This will require adequate drainage, flood management and flood protection
   provisions.

The challenge will be how to cope with the above developments and develop and
manage the water management and flood protection schemes in a sustainable way.

Population growth

Basis for the water management requirements is the worlds' population, its growth
and its standard of living. The present worlds' population and a prognosis of the
population growth are shown in Figure 1. Of special interest in this figure is the

distinction in least developed countries, emerging developing countries and developed countries. The majority of the worlds' population lives in the emerging developing countries. This category comprises Asia (excluding Japan), Latin America, the Caribbean and some other small regions. From Figure 1 it can be derived that population growth will take place in the least developed countries and the emerging developing countries. In the developed countries a slight reduction of the population is expected.

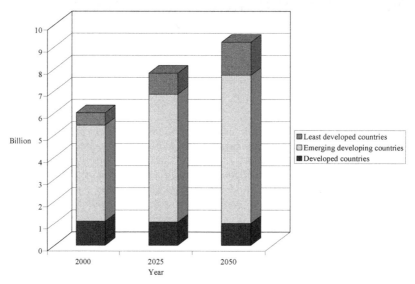

Figure 1. World population and growth in least developed countries, emerging developing countries and developed countries (after van Hofwegen and Svendsen, 2000).

Another interesting feature in the population growth is the migration for rural to urban areas. The expectation is that due to this the population in the rural areas in the least developed countries and the emerging developing countries will more or less stabilise and that the growth will be concentrated in the urban areas in these regions.

Water management for agriculture

In order to illustrate the different conditions under which agriculture can take place, I briefly like to give some data on water management related to agricultural production. With respect to this there are broadly speaking three agro-climatologic zone's, being: temperate humid zone, arid and semi arid zone and the humid tropical zone. In addition, in principle, four types of cultivation practices may be distinguished, being:
– rainfed cultivation, without or with a drainage system;
– irrigated cultivation, without or with a drainage system.

Dependent on the local conditions different types of water management with different levels of service will be appropriate (Schultz, 1993).

In the temperate humid zone agriculture generally takes place without a water management system, or with a drainage system only. Supplementary irrigation may be applied as well. In the arid and semi arid zone agriculture is normally impossible without an irrigation system. Drainage systems may be applied as well for salinity control and the prevention of water logging. In the humid tropical zone generally a distinction is made in cultivation during the wet and the dry monsoon. In many areas during the wet monsoon cultivation is possible with a drainage system only, although quite often irrigation is applied as well to overcome dry spells. In the dry monsoon irrigation is generally required to enable a good yield.

The total cultivated area on earth is about 1,500 million ha, which is 12% of the total land area. At about 1,100 million ha agricultural exploitation takes place without a water management system. Presently irrigation covers more than 260 million ha, i.e. 17% of world's arable land. Some characteristic figures of the ten countries with the largest irrigated area are given in Table 1. Irrigation is responsible for 40% of crop output. It uses about 70% of waters withdrawn from global river systems. About 60% of such waters are used consumptively, the rest returning to the river systems, in principal enabling its reuse downstream. Drainage of rainfed crops covers about 130 million ha, i.e. 9% of world's arable land. In about 60 million ha of the irrigated lands there is a drainage system as well. From the 130 million ha rainfed drained land it is roughly estimated that about 15% crop output is obtained. Some characteristic figures of the ten countries with the largest drained area are given in Table 2. In this table the total drained areas are given, while it is very difficult to differentiate between rainfed drained areas and drainage in irrigated areas.

Table 1. Some key figures for the ten countries with the largest irrigated area (International Commission on Irrigation and Drainage, 2000).

Country	Population in 10^6 in 1997	% of population in agriculture	Total area in 10^6 ha	Arable land in 10^6 ha in 1995	Irrigated area in 10^6 ha In 1996
India	960	61	329	170	57
China	1,243	68	960	96	50
USA	272	2	936	188	21
Pakistan	144	48	80	22	17
Iran	72	28	163	18	7
Mexico	94	23	195	27	6
Russia	148	11	1,171	208	5
Thailand	59	59	51	20	5
Indonesia	203	50	190	30	5
Turkey	63	48	77	27	4
Total	3,258		4,152	806	177
World	6,000		13,000	1,500	250

When we look at the actors in the field of agricultural water management the picture arises as shown in Figure 2. From this figure it can be derived that primary responsibility for agricultural water management rest with the governments, the irrigation and drainage agencies, either public or private and with the farmers. The many other involved parties play a contributing role. Their input is generally welcomed and required, but the final responsibility rests with the first three parties each with their own share and responsibilities.

Table 2.   Indicative key figures for the ten countries with the largest drained area (International Commission on Irrigation and Drainage, 2000, and CEMAGREF).

Country	Population in 10^6 in 1997	% of population in agriculture	Total area in 10^6 ha	Arable land in 10^6 ha in 1995	Drained area in 10^6 ha
USA	272	2	936	188	47.0
China	1,243	68	960	96	28.5
Indonesia	203	50	190	30	15.4
India	960	61	329	170	13.0
Canada	30	3	997	46	9.5
Brazil	163	19	851	66	8.0
Yugoslavia					5.8
Pakistan	144	48	80	22	5.7
Germany	82	3	36	12	4.9
Poland	39	23	32	15	4.2
Total					142.0
World	6,000		13,000	1,500	190

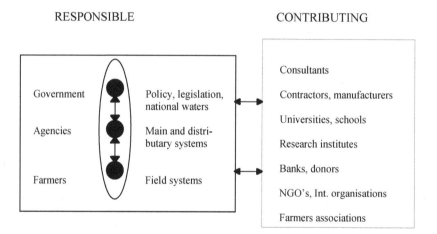

Figure 2.   Actors in the field of agricultural water management.

Water for food and rural development

Second World Water Forum

During the Second World Water Forum, which was held in March 2000 in The Hague, The Netherlands, the World Water Council (WWC) has presented a global 'Long Term Vision on Water, Life and the Environment in the 21st Century'. In the framework of the vision preparation process, among others, three major sector visions were prepared: 'Water for Food and Rural Development', 'Water for People' and 'Water and Nature'. The scope of these visions is 25 years. The International Commission on Irrigation and Drainage (ICID) has played a prominent role in the preparation of the sector vision on 'Water for Food and Rural Development' (van Hofwegen and Svendsen, 2000).

The sector vision of Water for Food and Rural Development indicates a required duplication in food production and gives general recommendations how this increase can be achieved. The major part of the increase in production would have to come from already cultivated land, among others, by water saving, improved irrigation and drainage practices, and increase in storages. It became fully clear during the vision preparation process that, especially in the least developed and emerging developing countries, huge efforts are required to:
– feed the still growing population;
– improve the standard of living in the rural area;
– develop and manage land and water in a sustainable way.

In order to achieve the required increase in food production in the framework of sustainable rural development, the following issues are generally considered to be of major importance related to dikes, dams and water management:
– availability of water and availability in space and time;
– need for increasing withdrawals with 15 - 20% to bridge mismatch between demand and supply in combination with water saving and improved efficiency in irrigation;
– need for increasing storages with 10 - 15%;
– basin wide planning for integrated development and management;
– inter basin transfers, shared rivers, conflict management;
– links between irrigation, drainage and flood protection, and food security, protection of the environment, sustainable rural development and livelihood;
– governance, legal, institutional and environmental issues.

Future directions

This brings us to the future directions. We cannot forecast these directions in detail, but tendencies can be observed that may sooner or later result in policy decisions, actual guidelines, or standards for design, implementation, operation, maintenance and management. These directions can be put under the following headings:
– integrated water management;
– developments in irrigation and drainage;
– integrated planning;
– sustainable development;
– acceptable environmental impacts.
I like to give some more background on each of these directions.

Integrated water management

For many centuries water management was mainly focused on water quantity control, by water supply or drainage. In most countries nowadays we may speak about water quantity and water quality control, although at different levels of service, more or less dependent on the respective standards of living. What we also see is that water management in many regions is becoming more adapted to diversification in land use, and not exclusively anymore only for agricultural use. In future most probably another step will be taken and we will come to an ecosystem approach.

Developments in irrigation and drainage

In the irrigation and drainage sector there are some specific issues that deserve our attention. Here we see that in the developed countries a lot has already been achieved, but that especially in the emerging developing countries these issues are far from being solved. It regards especially:
– required increase in efficiency and water saving;
– increased stakeholder participation;
– transfer of systems, or of responsibilities;
– modernisation;
– cost recovery.

Integrated planning

Irrigation and drainage are no isolated activities. They play a role in societies and have to be treated, also taking into account such issues. Therefore of importance are:
– links between irrigation, drainage and flood protection, and food security, rural development and livelihood;
– basin wide planning for integrated development and management.

Sustainable development

We are more and more concerned about the sustainability of our activities. In the past we did not have to bother so much about this, but increasing population pressure, changes in food production practices, and mining, or even exhaustion of resources have increased our concern. The following tendencies can be observed that will in different ways have an impact on agricultural water management:
– migration from rural to urban areas;
– requirement of higher yields per ha;
– increase in farm sizes, higher value crops, or part time farming;
– mechanisation in agriculture;
– increased application of fertiliser and pesticides.

I have already mentioned the expected increase in need for water for irrigation. However, although irrigation is the largest water user, it is not the strongest water user. Drinking water and industrial water supply are basically in a better position and already at a large scale water is transferred to these uses in several countries. This may not necessarily be a problem, while this water is generally coming back in the hydrological cycle. Problems arise when it comes back in a polluted form which constrains its reuse.

Acceptable environmental impacts

All water management projects have side effects. The challenge has been and will be to keep the negative environmental impacts at an acceptable level and to support

positive environmental impacts as far as reasonably possible. Of special importance related to water management are:
- controlled application of fertilisers and pesticides;
- quality criteria and quality control for drainage waters;
- prevention of water logging, salinization and mining of groundwater.

The debate on dams

At 16 November 2000 the report of the World Commission on Dams (WCD) was launched in London, Great Britain. The report got wide media attention. The report is one of the products in the global debate regarding the future development of dams and reservoirs. The debate is mainly initiated due to environmental concerns and resettlement issues. Extreme example of the problems related to dams is the Sardar Sarovar Dam in India where construction of the almost completed dam and related works was stopped for more than three years due to heavy opposition. A huge 150 km long irrigation canal that would have to take water from the reservoir was ready but could not be used. By October 2000 the Supreme Court of India gave its final judgement and stated that the project may be completed. Such cases are occurring more frequently nowadays, which will require a lot from future decision making.

In light of the global debate on dams and reservoirs ICID was asked to clarify its position. This has resulted in a 'Position paper on dams', which was almost unanimously approved by our National Committees during the International Executive Council meeting in Granada, Spain, September 1999 (International Commission on Irrigation and Drainage, 2000). As far as I can recall this was the first time that ICID took a position regarding a certain issue. Most probably ICID will make in the near future more of such position papers on topics that concern its National Committees. The position paper has been published and was presented during the Second World Water Forum and to the WCD. It is also available on ICID's website. At this place I would like to present to you one main statement of the position paper.

> 'Irrigation, drainage and flood control of agricultural lands are no longer options. They are necessary for feeding billions of people, for providing employment for rural poor and for protecting the environment. With respect to this ICID stresses that dams have played and will continue to play an important role in the development of water resources, especially in developing countries. A balance needs to be found between the requirements based on the needs of society, acceptable side effects and a sustainable environment.'

From ICID Granada Statement, 19 September 1999

Unfortunately it has not been possible to discuss the position paper with representatives of the WCD, nor was ICID consulted on the draft WCD report. When the report of the WCD was published ICID has disseminated it among its National Committees. The same was done by ICID's sister organisations, the International Commission on Large Dams (ICOLD) and the International Hydropower Association (IHA). There developed a great concern in these organisations about the WCD report, especially in the least and the emerging developing countries. This has, among others, resulted in a joint letter of the presidents of the three organisations addressed to the president of The World Bank. In the letter with attached comments it was stated that:

- the organisations consider the WCD report as a useful document to generate discussion, but absolutely inadequate, as it stands, to find the required sustainable solutions;
- the organisations do not accept the unbalanced judgement of the role of existing dams;
- the 26 guidelines as they currently stand are considered unrealistic for application.

I expect that this implies that for the above-mentioned organisations the WCD report is a closed book and that they will concentrate on new developments. With respect to this I consider the following points of importance:

- don't dispute the responsibility of the governments that have the responsibility for decision making on major water management and flood protection projects;
- promote co-operation to improve decision preparation and decision making processes;
- promote integrated approaches and sustainable solutions, to reduce the risk of too much focus on short term benefits;
- try to understand the tremendous challenges of the emerging developing countries.

Urban water management and flood protection

In urban and industrial water management and flood protection the roles are generally different than in agricultural water management. Again the government and the agencies will have their responsibility, but the local responsibility normally rests with the municipality and not with the individual citizens. The citizens may however be charged for at least a part of the costs of urban water management.

There is a very characteristic difference in design approaches between dams and dikes versus water management. In the design of dams and dikes the risk of loss of human lives is involved. In the design of urban drainage systems generally only the risk of damage to buildings and infrastructure plays a role.

In order to illustrate the differences that occur between the situation in the developing and the developed world I briefly like to present the conditions in Bangladesh and in The Netherlands. In Bangladesh there is a very rapid population growth and urbanisation going on, placing the government with the limited resources available for problems with an order of magnitude that it is almost impossible to find sustainable solutions. In The Netherlands urban water management and flood protection have been gradually developed during the centuries. Although these two countries just serve as examples, the general tendencies can be observed in many other countries as well.

Bangladesh

Some characteristic data for Bangladesh are:

- area 144,000 km^2, of which 95,600 km^2 is cultivable (67% of the land area);
- 120 million inhabitants, 815 per km^2, forecast for 2020 is 170 million;
- urban population: now 28 million, in 2020 about 80 million;
- rainfall 1,200 - 5,800 mm/year, annual average 2,300 mm/year;
- 8,613 km of embankments;

– a normal flooding, inundates about 27% of the cultivable area, 37% is inundated once every ten years. In 1998 60% of the country was inundated and floods stayed longer than two months.

What do we see in this example of Bangladesh, but also more in general in the least and emerging developing countries:
– rapid urbanisation;
– rapid increase in value of buildings and infrastructure in endiked areas, especially in the emerging developing countries;
– inadequate attention for economic optimal design standards for urban drainage, flood management and flood protection;
– increased flooding of urban areas resulting in increased damages;
– land subsidence, sea level rise, from drainage by gravity to drainage by pumping.

What can be done to prevent problems due to the above-mentioned developments as much as possible. I like to mention the following types of actions:
– promote creation of better linkages between urban water management and economic values;
– investigate present and (long term) future conditions during preparation and decision making for urban flood management and flood protection schemes;
– promote awareness among the urban population on their living conditions with respect to water management and flood protection.

The Netherlands

Some characteristic data of The Netherlands are:
– area 36,000 km^2, of which most is cultivable;
– 16 million inhabitants, 465 per km^2, very limited population growth;
– rainfall about 750 mm/year, spread more or less equally over the year;
– 20,000 km^2 endiked lands and drainage by pumping;
– development of water management over more than 1,000 years;
– urban population gradually growing;
– rapid growth of value of buildings and infrastructure in endiked areas.

If we look at the history of The Netherlands then we see that during the centuries a gradual transition has occurred from natural to cultivated land. The steps related to water management and flood protection can be summarised as follows:
– land drainage activities;
– artificial mounts;
– dikes, drainage and discharge sluices;
– pumping;
– water quality;
– integrated water management.
In Figure 3 the water management and flood protection measures, land subsidence and sea level rice are presented.

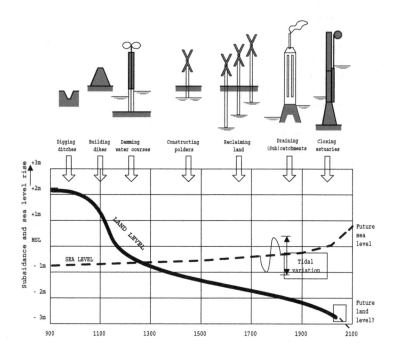

Figure 3.   Subsiding land and rising sea compel intervention.

At present an intensive debate is going on in The Netherlands on the future of its water management and flood protection. The activities and elements that play a role are as follows:
–   more room for the rivers;
–   sea level rise and land subsidence;
–   Water Management in the 21st Century;
–   long-term scenarios.
Room for the rivers is a policy where room will be given again for the rivers by placing certain sections of dikes more inland, deepening of riverbeds and floodplains and other measures where possible and appropriate.
    From these examples of Bangladesh and The Netherlands it can be seen in what completely different conditions the respective governments have to take their decisions. That such decisions will result in quite different solutions may be expected.

Concluding remarks

At the end of may presentation on dikes, dams and water management I like to present two concluding remarks:
–   as long as worlds population continues to grow and standards of living are improving the challenges increase to find sustainable solutions for water management and flood protection;

– dikes, dams and water management have to be developed and managed in an integrated way. Each time the balance has to be found between the needs of society, acceptable environmental impacts and a sustainable development.

References

CEMAGREF, *Data base on drainage*, Antony, France.

Hofwegen, P.J.M. van, and M. Svendsen, 2000, *A vision of water for food and rural development*, The Hague, The Netherlands.

International Commission on Irrigation and Drainage (ICID), 1999, *Role of dams for irrigation, drainage and flood control*, Position paper, New Delhi, India.

International Commission on Irrigation and Drainage (ICID), 2000, Draft *ICID strategy for implementing the sector vision on water for food and rural development*, New Delhi, India.

Schultz, E., 1993, *Land and Water Development. Finding a balance between implementation, management and sustainability*. Inaugural address delivered on the occasion of the public acceptance of the Chair of Land and Water Development at the International Institute for Infrastructural, Hydraulic and Environmental Engineering (IHE), Delft, The Netherlands.

World Commission on Dams (WCD), *Dams and development: a new framework for decision making*, Earthscan, London, Great Britain.

3

Living with dams

PROF. DR. H.L.F. SAEIJS AND KIRSTEN D. SCHUIJT

'The least we can do is to listen carefully,
and to learn from each other's mistakes'

The dams dilemma

Man has always tried to change his environment to serve his needs. Manipulation of natural systems has been practised on an ever-increasing scale, with ever increasing consequences for the characteristics and functioning of these systems. Many problems arose because man was not aware of, or neglected to take into account, the ecological laws. Manipulation of natural systems such as rivers, lakes and estuaries have included the construction of dikes and dams, cutting off river bends, and placing sluices and weirs. These interventions serve a wide variety of purposes for different stakeholders in society, such as industry, agriculture and civilians. Many interventions in water systems have proved to be an effective way to solve the actual problems of these different stakeholders. However, although the effects of interventions might be beneficial for one group of stakeholders, it is increasingly recognised that these interventions often also have negative effects on the functioning of the system, and thereby harm the interests of other stakeholders. The sum of the interventions might even (and often do) have a negative net total effect on the water system as a whole. One such intervention is the construction of *dams*. The number of large dams in the world is estimated at 45,000 and each year about 300 are added to this figure. The total amount of smaller dams in the word is estimated at 800,000!

Dams have convincingly proved their usefulness in preventing and mitigating floods and water scarcity and in the generation of electricity. They are indispensable in many present societies. But at the same time, they can lead to serious ecological, social and economic problems. The question that has to be answered is how this *'dams dilemma'* can be dealt with in a sustainable way.

The value of natural rivers

Value structurally underestimated

Dams are constructed in rivers. There, they are the direct or indirect cause of most of the problems they may provoke. As a result of the changing environmental conditions, landscapes will modify or even totally transform and all kinds of ecological, social and economic problems may develop. During the decision making process of dam construction the qualities and values of *natural rivers*[1] are structurally underestimated. The 'do nothing' alternative, as a consequence, is seldom (or never) a serious alternative (Saeijs, 1982). To reach a balanced decision, however, it is necessary to have enough knowledge about the features and potentials of the natural river, and about the new environment that is created as a result. Let's take a quick look at natural rivers. Rivers are an undeniable vital link in the hydrological cycle of water systems. Scientifically, there are three factors on which the existence of a river depends: availability of surface water, existence of a riverbed and an inclining earth surface. Rivers fulfil numerous important functions. The first and most vital function of a river is the *discharge of superfluous water* from the river basin along the surface of the river basin. A second important function is connected with this; the *erosion and the transport of sediments*, necessary for the downstream 'erosion/sedimentation equilibrium' in the delta of the river. For example, the Aswan dam in Egypt is the main cause of coastal erosion in the Nile delta. Another important function of a river is the distribution and long distance *transport of water and additives*.

Rivers are moreover environments with *unique life communities* in the water itself and on the floodplain. In this way, they contribute to global biodiversity. Furthermore, rivers produce products and services free of charge like fish, cleaning the water for drinking and irrigation, and providing space for human settlement and cultivation, unique for this type of environment (Constanza, et al, 1997).

Societies and ecosystems depend on these functions of rivers. The loss or degradation of these functions constitutes real costs to society. When they are lost, the replacement value of these kinds of products and services is tremendous. Think of the costs we have to pay to clean our water or to build dikes for safety reasons when these natural functions of rivers have disappeared.

An example of the cost associated with the loss or degradation of natural river functions can be seen in the Rhine river basin (Schuijt, 2001). The Rhine has been transformed from a natural meandering river containing numerous important benefits to society into a man-made river deprived of most of its natural functions. These natural functions constitute real economic benefits to society. As a result of human interventions, natural river functions like clean drinking water provision, fish production, nature, and natural retention capacity have more or less disappeared. The loss of these natural river functions are major costs to society. For example, the loss of clean drinking water provision amounts to US$ 663 million per year; the loss of nature equals US$ 640 million per year; and the loss of the natural retention capacity function of the Rhine results in costs of US$ 500 million per year. In the long run, these costs are borne by society.

It is important to realise that whatever interventions are conducted in a river basin, the river has to fulfil its functions at least at minimum level. In other words, after interventions have taken place, the river must contain at least the following essential features: enough water, sufficient dynamics (not too much, but certainly not too little), resilience, and connections between the subsystems. The importance

---

[1]   A natural river is a water course. that originated without the help of man, in which the water in a self made channel cleared a way from the higher parts of the earth surface to the lower parts, after which it mostly rushes into the sea or in a dry land area , where it by evaporation is drying up.

of dynamics for rivers is illustrated by the following example.

The transport capacity of a river is directly proportional to the sixth power of the rate of flow. A changing rate of flow, for instance as a result of dam construction, may have significant impact on the behaviour of sediments. Lowering the rate of flow then results in too much sediment upstream and too little downstream from the dam. An example of these system dynamics is the regularly occurring floods in the Nile delta during the months July, August and September. The construction of the Aswan dam abruptly ended these high discharges. However, the dam also had enormous impacts on the economically highly significant sardine population in the Mediterranean Sea. In fact, the disappearing dynamics affected the entire fishery sector in the Mediterranean area, leading to an ecological and economic disaster (Saeijs, 1982).

Ecosystem oriented cost benefit analyses

'The natural environment is almost universally undervalued in decision making and practically nowhere is there an 'awareness of ecological costs'. The loss of natural functions as a result of hydraulic engineering projects are important costs to society that should be included in the decision making process of a project. Since these natural functions are, however, largely outside the market system, they are often excluded from such decision making tools like cost-benefit analyses. When ecosystem functions are not incorporated in decision making, this leads to allocations that are economically inefficient. Although individual actors reap the benefits of the project, in the long run costs are borne by society as a whole. It is therefore vital to recognise the importance of natural ecosystem functions in decision making'

Bouma and Saeijs, 2000.

The incorporation of ecosystem functions in decision making can be achieved by valuing ecosystem goods and services into monetary terms (Constanza et al., 1997). Once monetary measures are found, these goods and services may be incorporated in a cost-benefit analysis, resulting in what is called an *ecosystem-oriented cost-benefit analysis*. In this way, the benefits of dam construction can be weighed against the costs (construction costs and costs of affecting the ecosystem) so that economically more efficient decisions can be made.

The loss of estuarine environments in the deltas of the Rhine, Meuse and Scheldt is a good example of the effects of excluding natural environments in decision making processes (Saeijs, 1999). Of the original 8,660 km^2 of estuaries in this delta in 1900, there remain only 3,930 km^2 in 2000: more than 54% of estuarine environment (4,730 km²) has disappeared within one century. When Costanza's key figures (1997) are applied to these estuaries and the new systems, the Gross National Nature Product of the estuaries in 1900 is estimated to have been about US$ 16 billion per annum^2. These water systems would presently represent a capital value of about US$ 336 billion. Of course, the figures are not absolute, but indicative. However, the message the figures convey is clear. Taking in account the gains of the new land and lakes, the loss in national nature product amounts to US$ 8.8 billion per annum while the net production loss can be estimated at US$ 6.6 billion. The net

---

[2]   One US$ was about NLG 2.40 (Dutch Guilders).

loss in capital is estimated US$ 138 billion. These costs of estuarine destruction have never been included in decision making tools like cost-benefit analyses during projects like the so-called Zuyderzee project and the Delta project. Current policy is aimed at (where possible) restoring estuarine environments. This too will cost a fortune.

Decision making in water management is quite frequently dictated by disasters. The challenge is to meet decisions based on rational arguments including long-term cost-benefit analyses to avoid disappointments. The time has come for economists and ecologists to work together. This will certainly result in new instruments for ecosystem management and perhaps in new applications of eco-economics.

Human interventions in river systems

The sum of 'the whole' is more than the sum of the elements

Constructing and managing dams cannot be seen as isolated activities (Figure 1): the sum of 'the whole' is more than the sum of the elements (Saeijs, van Westen and Winnubst, 1995). As a result of the need for e.g. safety, new land, cleaning wastes, and navigation, the impact on river systems by men is in most rivers tremendous. Each stakeholder of a river basin has his own priorities and seldom does he look over the boarders of his own sector to the impact on other activities in the entire river basin, or at the interrelationships between activities, or to the long-term impact of his activities.

The negative effects of dams that are becoming increasingly recognised are often the result of century-long activities in the wrong direction. For example, the cause of ever increasing water levels in The Netherlands is a result of 500 years of land reclamation in the floodplain and unsustainable use of the rivers' resources.

Dam impacts

There are many different dams:
– distinguished to used materials: earth fill-, rock fill-, and concrete dams;
– distinguished to environment: dams in mountains, on plains, in lower parts of the river or delta areas and, in estuaries;
– distinguished to main objectives for the construction of dams: hydropower, water storage, safety or land reclamation.

Evident is that the role of so many different dams cannot be caught in one single statement. Each group of dams needs its own attention: 'think global, act local' (Cosgrove and Rijsberman, 2000).

Dams have both benefits and concerns (World Commission on Dams, 2000). The main *benefits* of dams include the mitigation of floods, water supply for human needs and crops, navigation during periods of drought, recreation, recharge of ground-

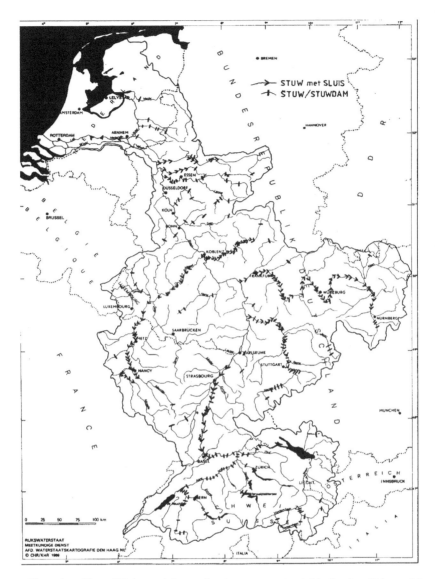

Figure 1.   About 480 dams, sluices and weirs were constructed in the river Rhine and its tributaries. The impact of all the structures together is much more than of each structure alone. The sum of 'the whole' is more than the sum of the elements.

water basins, hydro electric power, and so on. The dominating *concerns* of dams are safety risks, displaced persons, illnesses and diseases, a wide variety of environmental problems, the ruining of groundwater by reservoirs, sedimentation upstream of the dam, downstream salt up, and erosion in deltas.

At times, the ecological problems dominate (such is the case in the Aral Sea), at other times social problems (the case of the Narmada dam in India) and economic problems (the case of the Ataturc dam, Turkey) may dominate. Turkey with more than 600 dams constructed in the last fifty years; a tremendous performance. For the next fifty years another 600 dams are planned. Most of the dams are constructed to generate hydropower. Economic arguments dominate in the decision making process, because the income and other benefits are vital for the economy and development of the country. But a high price is paid (and will be) as a result of the loss of river integrity.

In the decision making process it is important to realise that it is not the dam but the specific problem at hand that needs special attention. The dam is only an instrument in river (basin) management, nothing more, nothing less. *Therefore, the focus must initially be on the management of the water system.* Then, all available options for the water system that help solve the problem need to be assessed. Once a dam is chosen as a preferred alternative, the impact of the dam on society and the environment, including the challenges of the new environment that is created, must be estimated. After all after dam construction, management of the entire river basin must continue.

Contemporary approaches in water management

Integrated water management in The Netherlands

In the course of 2000 years, hydraulic engineering activities in The Netherlands changed from small-scale to large-scale, defensive to offensive, short-term to long-term, specific to multifunctional, conflicting to harmonious, and from stemming the tides to controlling them. The local coastal engineering measures in the 1st to 11th centuries turned into well-organised dike building programmes from the 12th century onwards. Land-reclamation from inland lakes was carried out since the 16th century, which changed into large-scale and complex transformations in the 20th century. A universal pattern developed, illustrating that wherever in the world authorities are dealing with water, they will sooner or later be confronted with the following coherent range:
- *the area* to be managed (river-basin, river, lake, etc.);
- *the interests* associated with this area;
- *the potentials* of the ecosystems involved;
- *the machinery* necessary to ensure people's behaviour (laws, etc.) and to control;
- *the processes* of the system (sluices, barrages, dams, pumps, models, etc.);
- *the organisation* responsible for functional management;
- *the financial means.*

Experiences in The Netherlands with two big civil engineering projects last century, the Zuyderzee project (safety, water storage and land reclamation) and the Delta project (safety, water storage) have resulted in a revolution in water management that is now known as an *'Integrated water system approach'* (Saeijs and Bannink, 1978, Saeijs 1982, 1983, 1986, 1987, 1988, 1999, Ministry of Transport, Public Works and Water Management, 1985; Duursma and Davoren, 1983; Saeijs and Logemann, 1990, Saeijs and Turkstra,1994). This new type of Dutch water management has had its impact on the conferences in Rio de Janeiro (1992) and Dublin.

The need for a radical change

In order to manage water successfully, the relationships between several factors must be taken into account. Integrated water management aims to manage water systems (or land systems where water is an essential part) together with the associated lake and riverbeds, banks and groundwater, as one complete unit in relation to human interests. Integrated water management is a major step towards radical change in the management of the world's water systems as it recognises that:

- the water system as a whole is of primary importance. This includes everything that is related to the system: water, lake- and riverbeds, banks, salt-marshes and mudflats in tidal systems, infrastructure functions of rivers, lakes, canals, dams, dikes, barrages and pumping stations, substances that are contained within the water, as well as the living creatures and communities;
- many interested parties are involved in water systems. However, all may place different, sometimes conflicting, demands on the system. Interests and possibilities must be weighed up in a balanced way, taking account of their interrelationships;
- water systems function as an entity. One system cannot fulfil all the demands of different subsystems and stakeholders at the same time. The coherence in diversity should be preserved in policy;
- the wishes expressed by society and the possibilities offered by individual systems should be brought into line, and although water systems are multifunctional, choices should be made;
- water, with everything in it, is a moving part of the landscape - here today, gone tomorrow - and subject to changing authorities. Intervention at one place may have far reaching consequences for quality and utilisation elsewhere;
- all aspects of water management are required to be included as part of a balanced decision making process, taking full account of the interrelationships involved. This concerns safety, agriculture, settlement, industry, electricity supply, service sectors, shipping industry, fisheries, recreation, landscape and nature;
- water should no longer be considered as merely a raw material or a way of transport, but acknowledge the importance of a properly functioning aquatic ecosystem. Quantity and quality should be seen as interrelated subjects, as are ground- and surface-water;
- main infrastructures (including the major inland freshwaters, salt coastal waters and the continental shelf) that are managed by the government should be distinguished from a regional infrastructure, managed by local authorities.

The key question today is whether sufficient use is being made of the possibilities that water, infrastructure and creative methods of dealing with water systems can offer. For example, in developed countries like The Netherlands there is more need for sustainable management and for small-scale specialised multi-functional engineering, than for new large-scale hydraulic engineering projects

In the past, emphasis was laid on water as a medium, its use as a raw material and as a transport route, and its protection against the harmful consequences of human activities. Sustainable management still involves distributing water and protecting water systems from human intervention. However, the development of water systems also deserves attention. The wishes of society and the possibilities offered by water systems can and should be harmonised.

Aims for water management in the 21st *century*

The World Water Counsel formulates the aims for water management for this century as follows (Cosgrove and Rijsberman, 2000):

> 'Water is life. Every human being, now and in the future, should have access to safe water for drinking, appropriate sanitation, and enough food and energy at reasonable cost. Providing adequate water to meet these basic needs must be done in an equitable manner that works in harmony with nature. For water is the basis for all living ecosystems and habitats and part of an immutable hydrological cycle that must be respected if the development of human activity and well being is to be sustainable'.

They continue:

> 'We are not achieving these goals today, and we are on a path leading to crisis and to future problems for a large part of humanity and many parts of the planet's ecosystems. Business as usual leads us on an unsustainable and inequitable path' (See also: Saeijs and van Berkel, 1995 and Saeijs and Korver, 1999).

As a basis for decision making there are many who see water as a resource only for human uses. In the new approach water is recognised too as a vital part of ecosystems, on which all nature services, products and functions necessary for life on earth depends. Others look at political and administrative boundaries and sovereignty as basis for decision making. In doing so, it is seldom conform to the interests in the entire river basin. A narrow-minded or fragmented approach can lead to costly damage and restoration projects.

A good example of a water system that has been viewed primarily as a system for human use throughout history is the Rhine river basin. It has been transformed into a ship channel within almost one century. It is bordered by dykes and about 480 locks, weirs and dams have been constructed in its river basin (Figure 1). The river has been shortened by 25%. The harnessing of the river with dikes, land reclamation, uncoordinated management of all civil engineering structures, together with climate change, has resulted in more and more 'flood waves'. As a result new measures need to be taken to enhance the safety level in the basin countries. These measures include, next to the reinforcement of dikes, 'declamation'[3] to create more room for the river and bringing back the ecological flood regulating services of the river. In this way, many billions of dollars are now spent on the restoration of the river. To give an indication of the dimensions of the costs of these river rehabilitation, US$ 1.8 billion is allocated for the 'Room for the Rhine' project, aimed at managing floods by bringing back the ecological functions of the Rhine delta (including the reservation of retention areas and giving the river more space for natural flooding). Furthermore, US$ 40 million has been spent between 1970 and 1990 for the construction of waste treatment plants along the Rhine as a result of heavy industrial pollution, while an additional US$ 300 million needs to be invested to clean up the most heavily polluted sediment of the Rhine delta. These investments prove that natural functions of rivers are not dispensable as we once thought.

---

[3] Declamation is the opposite of reclamation, to give land back to the floodplain of the river or to the sea.

The wishes of each group of stakeholders of a river basin are understandable, perhaps feasible, seen from the perspective of the single stakeholder. However, taking the demands of all the stakeholders as a whole often results in devastation of the river system and causes costly problems. It is obvious that there is a need for tuning the demands among the different stakeholders in river basins.

Achieving the aims of the World Water Counsel and achieving a sustainable future for river systems and river basins requires drastic changes in the way water is managed. 'An holistic systemic approach relying on integrated water resource management must replace the current fragmentation in managing water' (Cosgrove and Rijsberman, 2000). A holistic or integrated approach implies water management that aims at connecting water tables, shores, and groundwater with all the materials, life communities and processes, as well as involved interest groups and stakeholders, in a comprehensive and unified approach.

Globalisation of the new approach

Increased awareness of the natural environment and its endangered situation is one of the most important developments of the late twentieth century. The United Nations 'Declaration on the Environment' and the Club of Rome's message in the 'Limits to Growth' left their mark on our thinking in 1972, followed in 1987 by immediate and world-wide agreement on the g concept of *sustainable development* as propagated in the Brundtland Report of the United Nations entitled 'Our Common Future'. In 1992, the United Nations Conference on Environment and Development (UNCED) put the issue into a global perspective and drew up a comprehensive action program in Agenda 21. This program stated, among others, that in order to meet sustainable development objectives, one should try to strike a balance between water and other natural resources. The most important strategic principles formulated at the UNCED conference were:

- policy and management need an integrated approach at the level of an entire river basin;
- management of water resources needs to be developed within a total package of policy measures on human health, production, protection and distribution of food, prevention and solution of accidental events;
- environments need to be protected and natural resources conserved;
- water management requires an integrated approach, based on the awareness that water is an inextricable part of the ecosystem and that water is also a social and economic asset;
- priority must be given to (1) fulfilling basic human needs and (2) at the same time protecting the earth's ecosystems.

These five basic principals (main recommendations) should be constraints for dam building and river (basin) management.

Next to these basic principles a number of institutional principles are applied like; the cause principle; the polluter pays principle; the equality principle; the profit principle; the sovereignty principle; the intergeneration principle, and the precaution principle.

Dams and the management of ecosystems

Resilience of ecosystems

When damming parts of rivers, the determining abiotic circumstances of the river basin should not be altered beyond the natural fluctuations of the natural river basin.

When the effects (for example due to artificial reservoirs) remain within the limits of this natural *resilience*[4],, no harm is done to the natural system. In that way, irreversible effects on ecosystems, and thus the negative impacts on socio-economic and environmental aspects, are avoided.

The challenge for each *new project* is to find out what the resilience is and how not to go beyond the limits of that resilience. (It is, for example, important to realise that the resilience for abiotic and biotic changes fluctuates with the seasons). The challenge for each *existing project* is to investigate what should be corrected in order to restore a situation in which the effects of the reservoirs stay within the limits of the resilience of the natural ecosystem of the river basin. This provides a logical explanation of the concept of sustainable use. Disturbing a river basin by human activity should never have a greater impact than the natural ranges and frequencies of disturbances that the natural river basin has to deal with. The word 'natural' is important here: the disturbances caused by human abuse of the river basin in the past shouldn't be taken into account of course. Following this, the flow of a river should never be completely dammed, because making a stagnant lake of a stretch of river clearly goes beyond the borders of the resilience of a natural river. Only parts of the flow should be used for reservoirs made in bypasses of the main stream only. Dams may offer many advantages but as soon as they are constructed on such a scale that the resilience of the natural system is eroded, they cease to be sustainable. Construction of small-scale dams instead of large-scale dam building should therefore be considered as a viable option.

Directed ecosystem development

If transformation is unavoidable it must be directed in such a way as to permit eventual replacement by healthy ecosystems in an area where other but equally vital natural functions can be fulfilled. These kinds of systems and changes require special caution, but must always be geared to achieve the most beneficial combination of the functions of man and nature. According to this line of thinking, the prevention of human activity is not an objective and certainly not an objective in its own right. The basic premise must be that the functioning of an area must be tailored to the needs of society on the condition that the ecosystems involved can continue to function in a healthy way. The foregoing raises three points that require particular attention:

- man has the right to make use of an area and to direct developments in a desired direction. At the same time it forces man to take care of the process and the final result - healthy functioning resilient ecosystems. It makes little sense to deny or frustrate this right in its most essential form. The emphasis in man's concern must be on better and more aware preparations for and supervision of change, with more systematic underpinning and ecological feelings. Of course 'the profit principle' is applicable (the one who has the profit has to pay all the costs, including the extra management costs like dredging of upstream sedimentation);
- if we are to arrive at a better mode of decision making concerning dams in river or estuarine ecosystems, particularly with respect to the 'ecosystems right' to protection as a type of environment, we must fully appreciate the ecosystems' characteristics and potential. Perhaps we must also learn to value them more

---

[4]   Resilience is (in this article) the ability of a river (eco)system to recover quickly from a setback such as caused by construction of a dam, or even a disaster.

highly and more consciously and to reflect this in our administration, management, and use of them. All too often the river or estuary as environmental types are disrupted or destroyed with no realisation of the productivity, and variety of life and potentiality that has been lost in the process;
- if the decision is taken to intervene, the changes or transformation must be thoroughly prepared (e.g., by research), guided, and kept under review. These are at least as essential components of a large-scale hydraulic engineering project as the structure that is built.

In decisions on matters of this kind, not only the aspirations already present in society (i.e., the primary social interests) but also the (potential) opportunities must be taken into account (Figure 2). This means that decisions that will result in changes must not be taken unless the existing situation (the 'do nothing' alternative), including the planned improvements, is studied at each stage. The results of these studies must be incorporated at each stage of the decision making process. Before any decision with major consequences for the systems is taken, it must first be known which natural resources (capital) will be lost by the change and what will be gained. The studies must cover technical and socio-economic as well as ecological aspects. *The point is not to prevent change (even if this were possible) but to channel the process of change along the right lines.* This requires an integrated policy plan on land-use, spatial organisation, and management, to be implemented in stages and have flexibility as to new changes and unknown factors.

Eco-pragmatism[5] will become increasingly important. Perhaps it will become the paradigm in water management of this century (Saeijs, 1999). The management strategy of *'Directed Ecosystem Development'*[6] a way of adaptive environmental assessment, design and management spans the entire process starting with the drafting of a project, through its implementation, and ending with the follow-up review (Holling, 1980). In terms of concern for the environment, this implies the following:
- environmental aspects must be included right from the start of preparations for the formulation of policy and be given the same importance as economic and social considerations so that policy making can benefit from natural forces and even enhance them;
- the policy-design stage must comprise periods of intensive innovation followed by periods of consolidation;
- policy must be framed such that some advantage can be drawn from a growing pool of information on socio-economic and environmental effects
- research designed to yield information must form part of an integrated research plan;
- the review machinery and the control mechanism must also be an integral part of the policy design on a par with the other components, and not simply be tackled at the end, once everything is completed.

---

[5]   Eco-pragmatism is a pragmatic application of ecological knowledge in assessment, design   and management of ecosystems like river systems, to direct the development of the ecosystem in a desire direction.
[6]   Related concepts: Guided-, Planned-, Controlled- Ecosystem Development.

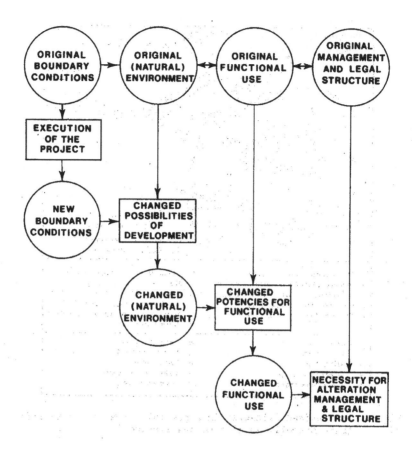

Figure 2. A chain of results as a consequence of a major human action in the environment (Saeijs and Bannink, 1978).

Challenges of modification and transformation processes

There is quite rightly a great deal of attention being devoted to threats and negative effects of dams. However, too little attention is being given to opportunities and possibilities of the new systems and their potential (Saeijs, 1978, 1982, 1994). Emphasis must be put on greater awareness of the many stakeholders involved, especially on the ecological implications of the project, and the necessity to incorporate *flexibility* that will make it possible to cope with changes and unexpected developments, and on recognition of the fact that processes are being dealt with. The alternative, comprising preservation of the existing situation (the 'do nothing' alternative), must play a much more important role in future preparations and decision making.

If a civil engineering structure is inevitable, the process of modification[7] and transformation[8] undergone by an area as a result of hydraulic engineering works must be considered at least as worthy of attention as the process of designing and constructing the works themselves. Both processes must play an important part in decision making, from the preparatory stage to the after-care stage.

Furthermore, a sectored approach must be avoided. Every effort must be made to achieve an integrated approach, taking the basin as a whole as the model. This demands smooth administrative co-operation. Multifunctional considerations should be given greater weight in the design of hydraulic projects. In this connection, more attention should be given to their function as an eco-technical management tool. Wide freedom of management must be incorporated into the design, to offer greater flexibility in response to changes, unanticipated events, new views on management; and such hydrodynamic works must be seen as subsidiary (as regulatory instrument) to the ecological and social functioning of the systems they can exert an influence on. A probabilistic approach to the design must therefore be related not only to the primary functions and the existence of the construction but also to the (future) requirements for modification, transformation, and management of the region that is affected. Decisions to execute hydraulic projects in regions for which (future) management plans do not exist must therefore be considered premature. The objective is not to resist change but to guide it properly. Ways to achieve these goals are indicated in the thesis 'Changing estuaries' (Saeijs, 1982).

Also the learning process is receiving too little attention both within the project and between dam projects. International organisations like ICOLD and the World Commission on Dams should promote examples of dams that contribute to sustainable development. Examples of (existing and planned) projects, which are not contributing to sustainable development, should be mentioned to the world explicitly.

Applying the principals in practice

Integrated river basin planning

There is an urgent need for *management plans* at the level of entire river basins. Sustainable development, management and use of natural (water) resources require integrated river basin planning. Dams could play a role in these plans, but they are not the first essential matter of concern in relation to sustainable development. In Europe the water policy of the European Commission is evolving from individual guidelines concerning different aspects (like water quality standards, pollution control, swimming water etc.) to a more integrated framework directive on water management. An important basic principle within the guideline is the organisation of water management at the level of an entire river basin. River basin authorities should propose and implement (not only with respect to water quality, but also water quantity issues) action programmes in order to solve the problems.

---

[7]   For example an estuary (Eastern Scheldt in The Netherlands) is modified from an open into a controlled estuarine environment with a reduced tide as result of a storm surge barrier.
[8]   For example an estuary (Grevelingen, The Netherlands) is transformed into a saltwater lake.

The long-term conservation of natural resources and the services they deliver to humankind (is a main objective of river basin plans so that a multiple and wise use of natural resources can be safeguarded for us and future generations. Many projects are carried out to serve local, regional or national needs without taking into account the real causes of problems and the effects of these projects on the entire river basin. Wide-ranging consideration must be given to the question of how the river basins should be managed internationally in the next century. Political decisions about water management issues may be motivated by short-term strategies but based primarily on an explicit long-term strategy. Therefore, an integrated river basin vision needs to be formulated for every river basin on earth.

Organisational arrangements

Management of water systems should also be organised at the entire river basin level. The basis should be laid by an international (holistic) evaluation study, followed by a policy analysis at the river basin level. In order to activate this international management of the river system, a step-by-step approach is perhaps the correct method:
1. development of a Co-ordinating Committee;
2. set-up a River Basin Commission;
3. when the time comes for this, set-up a Management Authority, with appropriate powers. The task package of this management authority might include: quantitative and qualitative water management; environmental protection; integrated management of existing infrastructure; ecological recovery of the river system; and co-ordination and harmonisation of new infrastructure construction.

Translation into a river basin action plan

In order to come to an integrated river basin action plan, five important steps should be undertaken:
–   as a first step, a clear description should be made of the state of the art in a river basin. What are the general features, what are the ecological and economical characteristics, what are the human interests? What are the borders of the natural resilience of the river basin? What sort of dams and other infrastructure are present in the river basin;
–   secondly, a problem definition should be made at the level of an entire river basin, including five major area of concern: protection from flooding (1), transport (2), energy demand (3), water availability and distribution (4) and maintenance of and ecological services. The abiotic factors are the basis for the that is present. This will continue as long as there is abiotic diversity. Or as long as there is no substantive change to it. Within the resilience of the ecosystem they can survive without permanent injury. (5). Socio-economic and ecological problems should be identified. Key-factors in inhibiting sustainable development should be identified;
–   thirdly, a long-term cost-benefit analysis, based on sustainable development of natural resources, should be set up for the entire river basin in its present state. Also long term cost-benefit analyses should be made for proposed solutions (step 4);
–   fourthly an inventory should be carried out of possible solutions to solve the problems. Active participation of stakeholders is highly desirable. In this respect

it is very important to consider a (large) dam or a series of dams as one of the possible instruments/alternatives for solving (social) problems, rather than as an objective in itself. When considering a dam project, the long-term costs to society and the environment should be studied and compared with these of alternative solutions;
– finally, recommendations should be formulated for a sustainable management approach for the entire river basin. A river basin action programme should be formulated and approved by the governments and stakeholders involved.

Conclusions and recommendations

The conclusions of this article can be summarised in the following main points:

A need for change

We are facing a dams dilemma. Business as usual leads us on an unsustainable and inequitable path; drastic changes in approaches and attitudes towards dams and river basin management are inevitable. Water is the basis for all living ecosystems and habitats and part of an immutable hydrological cycle that must be respected if the development of human activity and well-being is to be sustainable. Societies need to provide space for opportunities to solve the problems they face in other ways than only through the construction of dams in river basins

Ecosystems in decision making

In the decision making process the qualities and values of natural rivers are structurally underestimated. The 'do nothing' alternative, as a consequence, is seldom (or never) a serious alternative Dams are constructed in rivers, where they are the direct or indirect cause of problems, but also challenges. Societies depend on the proper functioning of rivers. Therefore, the loss or degradation of river functions constitutes real and high costs to society.

An integrated approach

A sectoral (fragmented) approach towards water management can lead to costly damage and expensive restoration projects. Therefore, an integrated approach towards water management is required. Integrated water management is based on the awareness that water is an inextricable part of the ecosystem and that water is also a social and economic asset. Human interventions in river systems, like dam building, must take into account the relationship between all the different interventions. Constructing and managing dams cannot be seen as isolated activities: the sum of the whole of actions is more than the sum of the elements. The effects of interventions must be assessed at the scale of the entire river basin and over the long term. Sustainable development, management and use of natural (water) resources require integrated river basin planning, and the wishes of society should be attuned to the possibilities of the ecosystem.

Focus first of all on the problem, not on the dam

The starting point in water management is the entire water system. The water system creates the conditions for solutions, which may or may not include a dam. The dam must therefore be seen as an instrument in river (basin) management only, nothing more, and nothing less. This should be at the basis of all decision making processes: the focus should be on the management of the water system and on the problems to solve.

Organisation at the level of a river basin

Organisation of water management should enclose an entire river basin. River basin authorities should propose and implement action programmes in order to solve the problems. The long-term conservation of natural resources and the services they deliver to humankind (e.g. productivity, water retention, energy, clean water for all kinds of purposes, biodiversity) is the main objective of river basin plans, in order to safeguard a multiple and wise use for us and future generations.

A vision and a river basin plan

There is first and foremost a need for a long-term vision and short-term agreements in river basin management. Water must become a structured element in spatial planning. Constructive co-operation and co-ordination of water policy in the river basin area is absolutely essential. Furthermore, there is an urgent need for management plans at the level of entire river basins. Dams could play a role in these plans, but they are not the first essential matter of concern in relation to sustainable development - this is the river basin

References

Bouma, J. and H.L.F. Saeijs, 2000, Eco-centric cost benefit analysis for hydraulic engineering in river basins. *New approaches to river management* ed. A.J.M. Smits, P.H. Nienhuis and R.S.E.W. Leuven. Backhuys Publishers, Leiden, The Netherlands.

Constanza, et al., 1997, The Value of the world's ecosystem services and natural capital, *Nature*, vol. 387 (15): 253-260.

Cosgrove, W.J. and F.R. Rijsberman, 2000, *World water Vision.* Making water every bodies business. World Water Council.

Holling, C.S., 1980, Adaptive environmental assessment. *Int. series on applied system analyses.3.* John Wiley and Son. New York, USA.

Ministry Public Works and Watermanagement, 1985, *Living with Water* Published by the Ministry, The Hague, The Netherlands.

Saeijs, H. L. F., 1982, Changing estuaries; a review and new strategy for management and design in coastal engineering. *Communications* 32. Directorate-General of Public Works end Water Management, The Hague, The Netherlands.

Saeijs, H.L.F., 1986, Towards control of en estuary. *Proceedings 4th International Conference On River Basin Management.* Sao Paulo, Brazil, 13-15 August 1986. Pub. ANAIS: 169-188.

Saeijs, H. L. F., 1987, Integrale Wasserwirtschaft. Ein neues Bewirtschaftungskonzept für die Niederlande. IAWR *Internationale Arbeitsgemeinschaft der Wasserwerke im Rheineinzugsgebiet. Kongress,* 11. Arbeitstagung 20-23 Oktober 1987, Noordwijk Proc. 40-76.

Saeijs, H.L.F., 1988, From treating of symptoms towards a controlled ecosystem management in the Dutch Delta. Lessons learned from 2000 years of living with water and eco-technology in The Netherlands. *Seminar 'The Dutch Delta,* 17 November 1988, on occasion of the Australian Bicentennial, Sydney, Australia. The Institute of Engineers Australia. Pub. Directorate-General of Public Works and Water Management, The Hague, The Netherlands.

Saeijs, H.L.F., 1992 and 1994, Creative in a changing delta. Towards a controlled ecosystem management in The Netherlands. *Proceedings IABSE 14th Congress,* 1 - 6 March 1992, New Delhi, India. Post Congress Report 97 - 109. And Proceedings, 18th ICOLD-Congress, 6 - 11 November 1994, Durban, South-Africa. Question 69; R.25: 371 - 395.

Saeijs, H.L.F., 1999, Levend water is goud waard. (Living water is worth gold) *Proceedings Symposium Het Verborgen Vermogen,* 26 march 1999, (at the occasion of the leave of Prof. Dr. H.L.F. Saeijs as Chief Engineer Director of the Directorate Rijkswaterstaat, Zeeland). Publication Rijkswaterstaat, directorate Zeeland. Middelburg, The Netherlands *(in Dutch).*

Saeijs, H. L.F. and B.A. Bannink, 1978, Environmental considerations in a coastal engineering project. *Hydrobiological Bulletin* 12: 314, 178 - 202.

Saeijs, H.L.F. and van M.J. Berkel, 1995, Global water crisis. The major issue of the 21st century; a growing and explosive problem. *Lecture on the Symposium 'BOTH SIDES OF THE DAM',* Organised by Technical University of Delft and NOVIB, 22 February 1995, Delft,. European Water Pollution Control. Publication Elsevier. Vol. 5,. 4: 26 - 40.

Saeijs, H.L.F., E.K. Duursma and W.T. Davoren, 1983, Integration of ecology in coastal engineering. Wat. Sci. Techn. 16: 745 - 757.

Saeijs, H.L.F., I.A. Flameling and L.A. Adriaanse, 1999, Eco-pragmatisme, Omgaan met rivieren, delta's, kust en zee in de 21ste eeuw. Boek: *'De Staat van Water'.* Opstellen over juridische, technische, financiele en politiek-bestuurlijke aspecten van waterbeheer. Eds. A van Hall, Th.G. Drupsteen and H.J.M. Havekes Uitg. Vermande: pp 29 - 42 *(in Dutch).*

Saeijs, H.L.F. and L. Korver-Alzerda, 1999, Coping with a World Water Crisis. Managing water in and around the city: an ecological approach. 34th International Planning Congress ISOCARP/AIU. 26th September - 2nd October 1998. Ponto-Delgada, Azores. Theme; 'Land and Water. Integrated planning for a sustainable future' (Keynote speech) *Proceedings/Final Report* 15 - 35.

Saeijs, H.L.F. and D. Logemann, 1990, Life history of a river basin; towards sustainable development of the Rhine catchment area. *Proceedings Congress 'Der Rhein, zustand und zukunft 'World Wildlife Fund, Tagungsbericht* 5:12 - 50.

Saeijs, H.L.F. and E. Turkstra, 1994, *Towards a Pan-European integrated river basin approach plea for a sustainable development of European river basins. European Water Pollution Control,* 4 (3): 16 - 28.

Saeijs, H.L.F. and M.J. van Berkel, 1995, Global Water crisis: The major issue of the 21st century; a growing end explosive problem. *European Water Pollution Control.* 5-.(4), 26 - 40.

Saeijs, H.L.F., C.J. van Westen and M.H. Winnubst, 1995, Time for revitalisation of the Rhine. Symposium ''The need for water. Storing water in riverbasins' July 6,1995, Oslo, Norway. During the executive Meeting of ICOLD, 2 - 8 July 1995, Oslo. *Proceedings Reservoirs in river basin development.* Santbergen and van Westen (eds.). Balkema, Rotterdam. ISBN 90 5410 5593 p.p. 3 - 24.

Schuijt, K.D., 2001, *'The Economic Value of Lost Natural Functions of the Rhine River Basin - Costs of Human Development of the Rhine River Basin Ecosystem',* Publikatiereeks nr. 36, Erasmus Studiecentrum voor Milieukunde, Erasmus University, Rotterdam, The Netherlands.

World Commission on Dams, 2000, Dams and development. *Earthscan Publications* Ltd London and Sterling VA.

4

Safe dams and dikes, how safe?

PROF. IR. J.K. VRIJLING

Introduction

Over the centuries all human civilisations have been threatened by natural hazards like hurricanes, floods, droughts, earthquakes, etc, that claimed the lives of individuals or entire groups bound by their residence or profession. Many activities have been deployed to protect man against these hazards. Even today money is spent to avoid or prevent natural hazards, because the consequences in developed societies have increased considerably. Other more recent hazards are man-made and result from the technological progress in transport, civil, chemical and energy engineering. One of the tasks of human civilisations is to protect individual members and groups against natural and man-made hazards to a certain extent. The extent of the protection was in historic cases mostly decided after the occurrence of the hazard had shown the consequences. The modern approach aims to give protection when the risks are felt to be high. This gives rise to the rather novel idea of acceptable risk.

As long as the modern approach is not firmly embedded in society, the idea of acceptable risk or safety may, just as in the old days, be quite suddenly influenced by a single spectacular accident as the catastrophe at Chernobyl, the plane crash at Schiphol airport in 1992, or even by non-calamitous threats like the Dutch river floods of 1993 and 1995. Here the political process is at work and public opinion is influenced not only by the accident itself, but also by the attention paid to it by the media and the politicians.

However according to the modern approach the politicians in an advanced technological society should not base their decisions to provide protection fully upon the above mentioned subjective and historical ideas of acceptable risk, but also use the outcome of risk analyses and probabilistic computations as a more objective basis. As the notion of probability of failure and the consequent risk forms the basis for the design of many technological systems, from simple river levees via multi-purpose dams to advanced jumbo-passenger-jets, that contribute to the welfare of modern nations, politicians should have an objective set of rules for the evaluation of risk. This paper proposes a possible set of ideas that may serve as a rational and more objective basis for technological design.

Historical development of dikes and dams

Although many contemporary academics would like to let us believe otherwise, the construction and the maintenance of dikes and dams has in most cases greatly benefited mankind. This cannot be refuted by pointing to the few cases were the engineering predictions have not become completely true or where unforeseen natural effects have reduced the benefits. A short sketch of the historical development will show why mankind built dikes and dams and still intends to maintain and construct them in future to some considerable extent. Around 0 AD the people in the delta of the Meuse and the Rhine lived on fertile but marshy soil that was regularly flooded by river floods and during winter storms by seawater. Especially the salty and unexpectedly rising storm surges threatened not only the fertility of the land but also human lives, livestock and housing. To protect their families, their cattle and their houses the people started to erect clay mounds to live on. This proved to be a beneficial solution. However the growing population and the secular rise of the sea level made the more frequent flooding of land and the consequent damage less acceptable. The far-reaching invention was to connect the mounds with dikes as flood protection for the cultivated areas. The wealth of the people grew after this considerable investment in technology. It is however frequently forgotten by engineers that this innovation could flourish only because governmental structures were established simultaneously. Where the erection and the maintenance of 'terps' (artificial mounds) could be organised within one family, the management of a dike system required the orderly and reliable co-operation of a group of families. To share the burden of investment and maintenance democratic water-boards were erected, that still exist today.

In the course of time the sea level kept rising and the land that was now gravity drained settled more than before. This forced the water-boards, after a period of implementing ever more sophisticated methods of draining by gravity, to install windmills. The necessity of this new investment was clear, as increasing numbers of the population had started to live beside the artificial mounds. The effectiveness of this rather costly new technological step forward showed itself in an increasing wealth of the ever-growing population. The improved artificial drainage increased however not only the agricultural yields, but also the settlement of the soil as is shown in Figure 1. The solution came with the advent of fossil fuels. The limits of wind power were overcome by installing coal and oil fired pumping stations that could be operated more precisely with respect time and level. The improved living conditions induced once again an increase of population numbers. The inhabitants of the polders showed their trust in the technology of dikes and drainage systems by building the required new towns preferably in the deepest parts of the polders. Even today only engineers ponder about the growing risk that results from the increased number of inhabitants of the deepest parts of the polders. The safety of the dikes is no matter of public discussion, also the incredible wealth and welfare is taken for granted; only the damage to 'nature' caused by dike maintenance is severely criticised.

For dams in rivers a similar history can be painted. A natural river flowing down a valley could sustain the living of a few people. The hydrological cycle of the river limited the population. The river threatened their livelihood by increased water levels and discharges during the wet season and by lack of water during the hot season. The first problem could be reduced by choosing higher grounds to settle, weighing the cost of transporting water against the risk of flooding. The second was less easily solved.

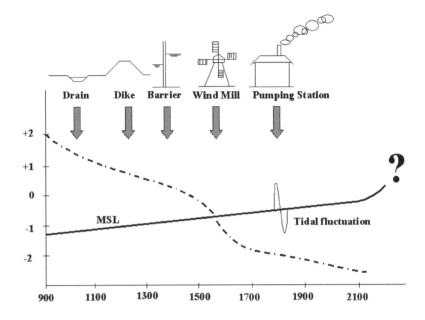

Figure 1. Rising sea and sinking land.

Damming the river and storing the water behind the dam for use during dry periods was a great invention, assuming that floods could be passed over or through the dam when needed. The increased reliability of the water supply over a longer part of the year improved agricultural yields considerably. Investments in irrigation systems could be defended on the basis of the more stable water supply. The healthy and growing population required and could afford to increase the capacity of the dam and the related storage. This was especially true, because the higher head improved the efficiency of power generation. The continuous availability of water and power formed the basis of growing wealth which stimulated a further increase in the number of people living downstream of the dam.

Moreover the increased capacity of the dam had also reduced the probability of flooding of the downstream area, thus adding to the availability of safe living space. In recent times the recreational possibilities of the reservoir may certainly have enhanced the attractiveness of living in the neighbourhood of an existing dam.

So for various reasons the population downstream of a dam has grown in living memory. These people seldom think of or discuss the remote possibility that the dam may fail, but enjoy its benefits. The remote prospect of failure is what makes some engineers in the ICOLD community nervous. This is caused by the enormous havoc that the flood wave will make in the densely populated valley downstream of the dam, not by their lack of trust in the safety of the dam. They know from engineering education and from the experience of many years of dam operation, that the probability of failure is less than 1/10,000 dam years. The ensuing catastrophe could however be beyond human imagination. The dam engineers would like to share the responsibility for the choice which risk is acceptable with the general public.

Risk of flood

Although the probability of failure of well-designed and constructed dams and dikes is very remote, the consequences can be extremely varied and large. Considerable numbers of people could lose their lives, most of the property in the threatened area will be damaged, income in the immediate future will be lost due to the incredible loss and disturbance, heritage and works of art will vanish and the environment will be threatened directly by the force of the water and indirectly by the release of toxic substances from chemical installations destroyed by the flood waters.

The risk consists of two components that together indicate the level of risk. The first is the probability of failure of the dam or dike. Failure may be caused by technical causes that are more or less familiar to dam- and dike engineers like overtopping, sliding, piping, etc. as sketched in the fault tree of Figure 2.

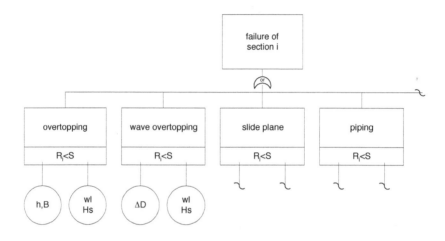

Figure 2.   A section of a dam or dike as a series system of failure modes.

Less obvious to classical engineers is however that management mistakes may also lead to failure. A small example for a sluice or a spillway gate is given in Figure 3. One should realise that generally the probability of human failure is larger than of technical failure. The total failure probability of the dike or dam system is found by combining all failure modes of all elements as shown in Table 1.

As pointed out the consequences of a flood are varied and large. To simplify this complicated picture of the consequence of such a disaster and to make it countable and open to quantitative analysis the losses are mostly schematised. In many practical cases the specified level of harm is limited exclusively to the loss of life N. In other cases the description of the damage is reduced to the counting of the material loss D in monetary units, in order to avoid the embarrassing discussion of the number of threatened population that might not survive a major failure (Dantzig et al., 1956). Either schematisation is not necessarily representative. Most probably society will look to the *total* damage caused by the occurrence of a hazard. This comprises the number of wounded as well as casualties, the material and economic damage as well as the loss of or harm to immaterial values. It is important to realise this fact when one discusses safety issues with the general public.

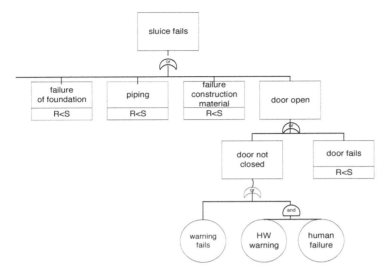

Figure 3.   The failure of a sluice caused by technical or human failure. A spillway gate could fail to open due to similar causes.

Table 1.   Calculation Table For The Overall Probability of Flooding of a Polder Defended By Dikes, A Dune And A Sluice (Vrijling, 2001).

Section	Overtopping	Piping	etc.	Total
Dike section 1.1	$p_{1.1}$ (overtopping)	$p_{1.1}$ (piping)	$p_{1.1}$ (etc.)	$P_{1.1}$ (all)
Dike section 1.2	$p_{1.2}$ (overtopping)	$p_{1.2}$ (piping)	$p_{1.2}$ (etc.)	$P_{1.2}$ (all)
etc.	..	..	..	..
Dune	p_{dune} (overtopping)	p_{dune} (piping)	p_{dune} (etc.)	p_{dune} (all)
Sluice	p_{sluice} (overtopping)	p_{sluice} (piping)	p_{sluice} (etc.)	p_{sluice} (all)
Total	p_{all} (overtopping)	p_{all} (piping)	p_{all} (etc.)	p_{all} (all)

The consequences in case of the flooding of polders or valleys will be estimated from two points of view. First an estimate of the probability to die for an individual residing at some place in the polder or valley will be given. The most practical form of presentation might be a contour plot of the risk as a function of the place in the polder or valley.

Secondly the total number of people that will drown in a flood must be estimated. If the specified level of harm is limited to the number of casualties the risk may be modelled by the probability density function (pdf) of the number of deaths. As the chance of failure of dams and dikes is generally very remote the pdf will show a large mass of $1 - p_f$ at the origin exemplifying the normal safe functioning without any problems. However to the right on the x-axis a part of the pdf with mass p will show the range of possible outcomes in terms of deaths if the dam or dike fails. Here

p_f is the probability of failure. The performance of an integration from the right will produce the probability of exceedance curve of the number of deaths. This curve starts at 1 at the vertical axis and drops immediately to p_f also called the FN-curve due to a specific hazard. The FN-curve is commonly depicted on double-log axes as shown in the graph at the bottom of Figure 4. Thirdly the consequence part of a risk may also be limited to the total material damage expressed in monetary terms. The graphical presentation could be the same as indicated in Figure 4 exchanging N for D at the horizontal axis. As pointed out by Vrijling (1997) it is useful and illustrative to calculate the expected value and the standard deviation of N and D beside the graphical representation by the pdf or the FN/D-curve.

In many practical cases the FN-curve is calculated numerically, leading to a stepwise decreasing function as given in Figure 5. In this graph the effect of two categories of measures to reduce the risk can be indicated. If the safety of the dam or the dike is increased, the graph will be lowered. To narrow the FN/D-curve the maximal consequences of a breach must be reduced. This seems only possible by spatial planning measures like the restriction of new settlements to relatively higher grounds. A difficult proposition as any planner knows.

It should be stressed again, however, that the reduction of the consequences of an accident to the number of casualties or the economic damage might not adequately model the public's perception of the potential loss. The aim of the schematisation is to clarify the reasoning at the cost of accuracy.

How safe should the dam or the dike be?

The most complex, controversial and sensitive issue that hugely matters to society is the relation between classical engineering safety as guided by experience and codes on the one hand and the philosophies about acceptable risk levels, that have been evolving in some societies in recent years. The first ideas were developed for certain industrial activities sited in the neighbourhood of housing and offices (Health and Safety Executive, 1989 and Ministry of Housing, Land Use Planning and Environment, 1988).

The question of acceptable risk is sometimes explicitly but mostly implicitly at the basis of every engineering design decision. The popular request for absolute safety is unattainable, because it would require the spending of an unlimited amount of society's resources. In the design of dams and dikes all uncertainties have to be exposed and included to ensure that the trade off between reduced uncertainty i.e. extra safety and the required increased use of society's resources is explicitly made. Clearly this involves a careful balancing of the costs to society of:
- the planning and design activities;
- the construction of the dam or dike;
- the area occupied by the structure;
- the reduction (or the increase) of environmental values;
  against the benefits:
- reliable water supply or living area safe from flooding;
- reduced probability of loss of life;
- reduced probability of material loss;
- reduced probability of damage to the environment by toxic or other releases;
- increased economic development and welfare.

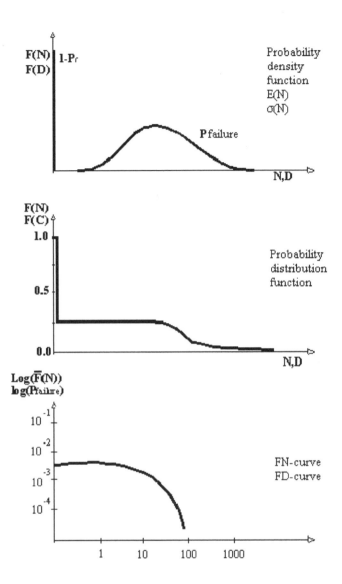

Figure 4.   Risk of flooding.

The well informed and scientifically based decision making that properly takes account of all uncertainties, costs and benefits is crucial in the process. Here guidance to arrive at a credible defensible and transparent risk analysis could be given, but the formulation of a standard for e.g. acceptable risk, is judged to be too difficult as it would require a generalisation of the cost/benefit decision for a range of hazardous activities. These decisions are related to the geographical situation, the economical development, the cultural values and the political system of each country. Here an exposition will be given of the current thinking in The Netherlands and some neighbouring countries.

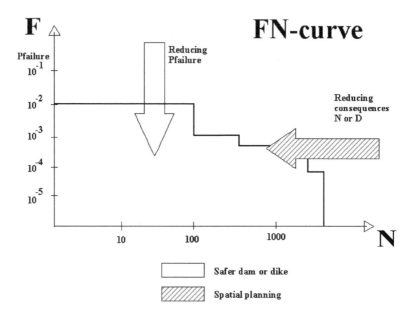

Figure 5.   A numerically calculated FN-Curve.

In most treatises of acceptable risk two positions are discerned. The point of view of the individual, who decides to undertake an activity weighing the risks against the direct and indirect personal benefits. And secondly the point of view of the society, considering if an activity is acceptable in terms of the risk-benefit trade off for the total population. The first point of view leads to the personally acceptable level of risk or the acceptable individual risk, defined in ICE as 'the frequency at which an individual may be expected to sustain a given level of harm from the realisation of specified hazards'. It was explained above that in many practical cases the specified level of harm is limited exclusively to the loss of life. Similarly the notion of risk in a societal context is reduced to the total number of casualties using a definition as in ICE: 'the relation between frequency and the number of people suffering from a specified level of harm in a given population from the realisation of specified hazards'. If the specified level of harm is limited in this way the societal risk may be modelled by the FN-curve due to a specific hazard. As stated above the consequence part of a risk may also be limited to the total material damage expressed as a FD-curve in monetary terms.

The smallest component of the social acceptance of risk is the assessment by the individual. Attempts to model this are not feasible; therefore it is proposed to look to the preferences revealed in the accident statistics. The fact, that the actual personal risk levels connected to various activities show statistical stability over the years and are approximately equal for the Western countries, indicates a consistent pattern of preferences. The probability of losing one's life in normal daily activities such as driving a car or working in a factory appears to be one or two orders of magnitude lower than the overall probability of dying. Only a purely voluntary activity such as mountaineering entails a higher risk (Figure 6).

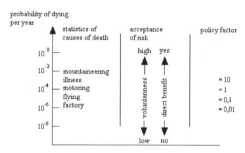

Figure 6.   Personal risks in Western countries, deduced from the statistics of causes of death and the number of participants per activity.

If one neglects a secular downward trend of the death risks due to technical progress, it seems permissible to use them as a basis for decisions with regard to the personally acceptable probability of failure of activity i in the following way:

$$P_{fi} = \frac{\beta_i \bullet 10^{-4}}{P_{d|fi}}$$

where $P_{d|fi}$ denotes the probability of being killed in the event of a failure. In this expression the policy factor β_i varies with the degree of voluntariness with which an activity i is undertaken and with the benefits perceived. This factor ranges from 100 in the case of complete freedom of choice like mountaineering, to 0.01 in case of an imposed risk without any perceived direct benefit, like the siting of a hazardous installation near a housing area. A proposal for the choice of the value of the policy factor β_i as a function of voluntariness and benefit is given in Table 2.

For the safety of dikes and dams a β_i -value of 1.0 to 0.1 is thought to be applicable.

The judgement of societal risk due to a certain activity should be made on a national level. The risk on a national level is the aggregate of the risks of local reservoirs or polders. Starting with a risk criterion on a national level one should evaluate the acceptable local risk level, in view of the actual number of reservoirs or polders, the cost/benefit aspects of the activity and the general progress in safety, in an iterative process with say a twenty to fifty year cycle.

Table 2.   The value of the policy factor β_i as a function of voluntariness and benefit.

β_i	Voluntariness	Direct benefit	Example
100	Voluntary	Direct benefit	Mountaineering
10	Voluntary	Direct benefit	Motor biking
1.0	Neutral	Direct benefit	Car driving
0.1	Involuntary	Some benefit	Factory
0.01	Involuntary	No benefit	LPG-station

The determination of the socially acceptable level of risk assumes also that the accident statistics reflect the result of a social process of risk appraisal and that a standard can be derived from them. The formula should account for well-known risk aversion in society. Relatively frequent small accidents are easily accepted, while one single rare accident with considerable consequences like a flood (or more recently the fireworks disaster in Enschede and the pub-fire in Volendam) fills the newspapers for days, although the expected number of casualties is equal for both cases. The standard deviation of the number of casualties, that is much larger for the second case, reflects this difference to some extent.

Risk aversion can be represented mathematically by adding the desired multiple k of the standard deviation to the mathematical expectation of the total number of deaths, $E(N_{di})$ before the situation is tested against the norm of β_i 100 casualties for The Netherlands:

$$E(N_{di}) + k \bullet \sigma(N_{di}) < \beta_i \bullet 100$$

where: k = risk aversion index

To determine the mathematical expectation and the standard deviation of the total number of deaths occurring annually in the context of activity i, it is necessary to take into account the number of independent places N_{Ai} where the activity under consideration is carried out.

The translation of the nationally acceptable level of risk to a risk criterion for one single installation or polder where an activity takes place depends on the number of casualties for accidents of the activity under consideration. In order to relate the new local risk criterion to the FN-curve, the following type is preferred:

$$1 - F_{N_{dp}}(x) < \frac{C_i}{x^2} \quad \text{for all } x \geq 10$$

The principle of the societal risk criterion limiting the risk of a single local installation is given in Figure 7.

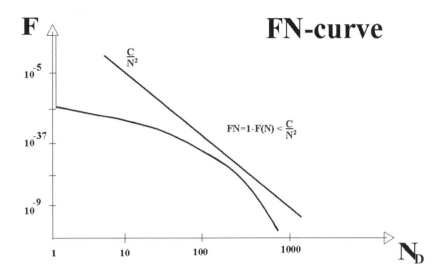

Figure 7.   Societal risk criterion for the FN-curve at local level.

If the expected value of the number of deaths is much smaller than its standard deviation, which is often true for the rare calamities studied here, the value of C_i reduces to:

$$C_i = \left[\frac{\beta_i \bullet 100}{k \bullet \sqrt{N_{A_i}}} \right]^2$$

The problem of the acceptable level of risk can be also formulated as an economic decision problem as explained earlier. The expenditure I for a safer system is equated with the gain made by the decreasing present value of the risk (Figure 8). The optimal level of safety indicated by P_f corresponds to the point of minimal cost.

$$min(Q) = min(I(P_f) + PV(P_f \bullet S))$$

where: Q   = total cost
              PV   = present value operator
              S   = total damage in case of failure

If despite ethical objections, the value of a human life is rated at s, the amount of damage is increased to:

$$P_{d\mid fi} \bullet N_{pi} \bullet s + S$$

where: N_{pi} = number of inhabitants in polder i.

This extension makes the optimal failure probability a decreasing function of the expected number of deaths. The valuation of human life is chosen as the present value of the nett national product per inhabitant. The advantage of taking the possible loss of lives into account in economic terms is that the safety measures are affordable in the context of the national income.

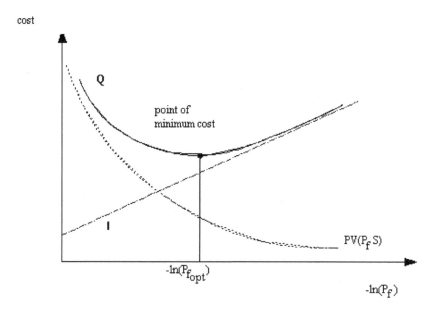

Figure 8.   The economically optimal probability of failure of a dam or dike.

In assessing the required safety of a dike system the three approaches described above should all be investigated and presented. The most stringent of the three criteria should be adopted as a basis for the 'technical' advice to the political decision process. However all information of the risk assessment should be available in the political process.

Conclusions

It is shown that contrary to what many contemporary academics would like to let us believe, the construction and the maintenance of dikes and dams has in most cases greatly benefited mankind. This cannot be refuted by pointing to the few cases were the engineering predictions have not become completely true or where unforeseen natural effects have reduced the benefits. A short sketch of the historical development showed that mankind built dikes and dams to improve living conditions, wealth and welfare. A consequence was the continuous growth of the population in polders and downstream valleys. It is reasonable to expect that mankind, especially in less developed areas, still intends to maintain and construct dams and dikes in future to some considerable extent.

The probabilistic approach has great advantages compared with the present classical engineering approach. The event that the dam or dike is meant to prevent (flooding), comes at the centre of the analysis. The contribution of all elements of the system and of all failure mechanisms of each element to the probability of flooding is calculated and presented. The possibility to include the probability of human failure in the analysis of the management of the structures is especially attractive and useful.

Finally an approach is sketched to define the level of acceptable risk. The decision on the level of acceptable risk is a cost/benefit judgement that must be made from individual as well as from societal point of view. A system of three rules is developed to support the decision how safe the dikes should be. The individual acceptable risk criterion, the societal acceptable risk criterion and the economical optimal societal risk should all be calculated for a specific project. The most stringent of the three criteria should be adopted as a basis for the 'technical' advice to the political decision process. However all information of the risk assessment should also be made available to the political process. A decision that is political in nature, must be made open and democratically, because many differing values have to be weighed. The economic optimisation may however show that the economic activity in the polders and the areas downstream of dams has grown so much since the start of the 20th century that a fundamental reassessment of the acceptability of the flood risks is justified. Moreover the image of the polders and valleys downstream dams as safe areas to live, work, and invest in is an important factor to consider especially when the population continues to grow and ever more ambitious private and public investments are planned.

References

Health and Safety Executive (HSE), 1989. *Risk criteria for land-use planning in the vicinity of major industrial hazards*, HM Stationery Office.

Ministry of Housing, *Land Use Planning and Environment* (VROM), 1985, LPG Integral Study (in Dutch). The Hague, The Netherlands.

CUR, 1988. *Probabilistic design of flood defences*. Gouda, The Netherlands.

Van de Kreeke, J. and A. Paape. On the optimum breakwater design, *Proceedings 9th IC Coastal Eng*.

Dantzig, V.D. and J. Kriens, 1960. The economic decision problem of safeguarding The Netherlands against floods. *Report of Delta Commission*, Part 3, Section II.2 (in Dutch). The Hague, The Netherlands.

Dantzig, V.D., 1956. Economic Decision Problems for Flood Prevention, *Econometrica* 24, pp 276 - 287, New Haven.

Ministry of Housing, *Land Use Planning and Environment* (VROM), 1992. Relating to risks (in Dutch). The Hague, The Netherlands.

Institute of Chemical Engineering (ICE), 1985. *Nomenclature for hazard and risk assessment in the process industries*, ISBN 85 295184 1.

Vrijling, J.K. et al., 1995. Framework for risk evaluation, *Journal Hazardous Materials* 43.

Vrijling, J.K. et al., 1998. Acceptable risk as a basis for design, *Journal Rel. Engin and System Safety* 59.

Vrijling, J.K. and P.H.A.J.M. van Gelder, 1998. *Societal risk and the concept of risk aversion*, pp 45 - 52, ESREL 1997.
Vrijling, J.K., 2001. Probabilistic design of water defence systems in The Netherlands, *Journal Rel. Engineering and System Safety* (to be published).

5

Dams and development

The report of the World Commission on Dams

Mr. Jamie Skinner

Why a World Commission on Dams?

The intensifying controversy surrounding large dams is not about the technical designs themselves, but their social and environmental consequences and the decision making processes that lead to their construction. It revolves around whether a dam is the most suitable option to meet a particular development need, and if so, the extent that its implementation can accommodate the increasing social and environmental concerns. At the heart of the dams debate are issues of equity, governance, justice and power - issues that underlie the many intractable problems faced by society. The World Commission on Dams (WCD) was established as a response to the rising conflicts over dam projects and provided a forum within which all sides could present their views.

During the 20th Century, large dams emerged as one of the most significant and visible tools for the management of water resources. The more than 45,000 large dams around the world have played an important role in helping communities and economies harness water resources for food production, energy generation, flood control and domestic use. Current estimates suggest that some 30 - 40% of irrigated land worldwide now relies on dams and that dams generate 19% of world electricity. While the immediate benefits were widely believed sufficient to justify the enormous investment made - total investment in large dams worldwide is estimated at more than US$ 2 trillion - secondary and tertiary benefits are also evident. These include food security, local employment and skills development, rural electrification and the expansion of physical and social infrastructure.

As experience has accumulated and better information on the performance and consequences of dams became available, opposition has grown. Global estimates of the magnitude of impacts include some 40 - 80 million people displaced by dams. Many more people's lives are affected but go unrecognised. Sixty percent of the world's rivers have been affected by dams and diversions. As a consequence, intense conflicts arose between governments building dams and people affected by dams. The continuing struggle against proposed dams on the Narmada river in India is an example and has resulted in costs to all involved - the affected communities, intended beneficiaries, and government alike.

In April 1997 in Gland, Switzerland, representatives of diverse interests were invited by the World Bank and The World Conservation Union (IUCN) to discuss the highly controversial issues associated with large dams. As a result, the WCD was established and started work in June 1998. The Commission's credibility rests on a process that incorporates diversity of perspectives at all levels, and on an inclusive, transparent and participatory work programme. The WCD has conducted the most comprehensive, global and independent review of large dams to date, and used this review as a basis for its recommendations. A Stakeholder Forum, of 68 members built on the participation at Gland, was created to interact closely with the WCD. The Forum was a unique feature of this Commission and ensured the inclusion of other partner's views and opinions directly into the debate. Funding was sought from a broad range of stakeholders and 53 public, private and civil society organisations provided untied financial support to the process.

How has the Commission addressed these issues?

The Commission has added a new body of knowledge to existing databases and information on large dams, looking at alternative ways of meeting irrigation, water supply, energy, flood management requirements, the impacts of dams and mitigation measures, and the processes of development planning. The Commission's objectives were to:
- review the development effectiveness of large dams and assess alternatives for water resources and energy development;
- develop internationally acceptable criteria, guidelines and standards, where appropriate, for the planning, design, appraisal, construction, operation, monitoring and decommissioning of dams.

The resulting Knowledge Base includes eight detailed case studies of large dams, country reviews for India and China, a briefing paper for Russia and the Newly Independent States, a Cross Check Survey of 125 existing dams, 17 thematic review papers, 130 contributing papers as well as the results of four regional consultations and more than 900 submissions to the Commission. The entire WCD Knowledge Base is available on its website (www.dams.org).

Undertaking a global policy process on this scale with the attendant diverging perspectives and controversies was a major challenge. There were many times when the Commission's balance was questioned from interested parties on both sides of the debate. The Commission recognised such concerns, was open to dialogue throughout, and embraced the controversial and diverging views it encountered as part of its approach to knowledge gathering. The WCD continued with the process of dialogue and negotiation, which culminated with its final report '*Dams and Development - a New Framework for Decision Making*' which was released in November 2000. The WCD process itself was studied as a potential model for other initiatives and will be analysed in a forthcoming report by the World Resources Institute.

Main findings of the WCD global review

Based on its Knowledge Base, the Commission found that:
- large dams display a high degree of variability in delivering predicted water and electricity services - and related social benefits - with a considerable portion

falling short of physical and economic targets, while many continue generating benefits beyond their projected economic life;
- large dams have demonstrated a marked tendency towards schedule delays and significant cost overruns;
- large dams designed to deliver irrigation services have typically fallen short of physical targets, did not recover their costs and have been less profitable in economic terms than expected;
- large hydropower dams tend to perform closer to, but still below, targets for power generation, generally meet their financial targets but demonstrate variable economic performance relative to targets, with a number of notable under- and over-performers;
- large dams generally have a range of extensive impacts on rivers, watersheds and aquatic ecosystems - these impacts are more negative than positive and, in many cases, have led to irreversible loss of species and ecosystems;
- efforts to date to counter the ecosystem impacts of large dams have met with limited success owing to the lack of attention to anticipating and avoiding impacts, the poor quality and uncertainty of predictions, the difficulty of coping with all impacts, and the only partial implementation and success of mitigation measures;
- pervasive and systematic failure to assess the range of potential negative impacts and implement adequate mitigation, resettlement and development programmes for the displaced, and the failure to account for the consequences of large dams for downstream livelihoods have led to the impoverishment and suffering of millions, giving rise to growing opposition to dams by affected communities world-wide;
- since the environmental and social costs of large dams have been poorly accounted for in economic terms, the true profitability of these schemes remains elusive.

The full Global Review documented by the Commission covers five chapters and 159 pages in the Report and provides an integrated insight into performance and impacts of dam projects. After extensive dialogue with those for and against large dams, the Commission believes that there can no longer be any justifiable doubt about five key points:
- dams have made an important and significant contribution to human development, and the benefits derived from them have been considerable;
- in too many cases an unacceptable and often unnecessary price has been paid to secure those benefits, especially in social and environmental terms, by people displaced, by communities downstream, by taxpayers and by the natural environment;
- lack of equity in the distribution of benefits has called into question the value of many dams in meeting water and energy development needs when compared with the alternatives;
- by bringing to the table all those whose rights are involved and who bear the risks associated with different options for water and energy resources development, the conditions for a positive resolution of competing interests and conflicts are created;
- negotiating outcomes will greatly improve the development effectiveness of water and energy projects by eliminating unfavourable projects at an early stage, and by offering as a choice only those options that key stakeholders agree represent the best ones to meet the needs in question.

The Commission's decision making framework

The Commission's final report presents a new framework for decision making that encompasses how to deal with the differing interests and conflicts over dam projects. It is based on five core values: equitability, sustainability, efficiency, participatory decision making, and accountability. The approach focuses on recognising rights and assessing risks as a means to identify legitimate stakeholders and negotiate development outcomes. The Report encourages a move beyond the prevailing 'balance-sheet' approach where one person's gains are traded off against another's loses, to decision making were social and environmental concerns are given a higher significance in a multi-criteria analysis. Seven strategic priorities, corresponding policy principles for water and energy resources development, and a series of practical criteria and guidelines are presented.

Clarifying the rights context for a proposed project is seen as an essential step in identifying the legitimate claims and entitlements that may be affected by the project or its alternatives. It is also a pre-condition for effective identification of legitimate stakeholder groups that are entitled to a formal role in the consultative process, and eventually in negotiating project-specific agreements relating, for example, to benefit sharing, resettlement and compensation.

The assessment of risk adds an important dimension to understanding how, and to what extent, a project may impact on people's rights and risks (both voluntary risk takers and involuntary risk bearers). In the past, many groups have not had an opportunity to participate in decisions that imply major risks for their lives and livelihoods, thus denying them a stake in the decision making process commensurate with their exposure to risk. Within this framework the Commission has developed seven strategic priorities and related policy principles, that provide a principle and practical way forward for decision making. The strategic priorities are presented in the style of intended outcomes. Each are supported by four to five policy principles that are detailed in the Report.

1. Gaining public acceptance

Public acceptance of key decisions is essential for equitable and sustainable water and energy resources development. Acceptance emerges from recognising rights, addressing risks, and safeguarding the entitlements of all groups of affected people, particularly indigenous and tribal peoples, women and other vulnerable groups. Decision making processes and mechanisms are used that enable informed participation by all groups of people, and result in the demonstrable acceptance of key decisions. Where projects affect indigenous and tribal peoples, such processes are guided by their free, prior and informed consent.

2. Comprehensive options assessment

Alternatives to dams do often exist. To explore these alternatives, needs for water, food and energy are assessed and objectives clearly defined. The appropriate development response is identified from a range of possible options. The selection is based on a comprehensive and participatory assessment of the full range of policy, institutional, and technical options. In the assessment process social and environmental aspects have the same significance as economic and financial factors. The options assessment process continues through all stages of planning, project development and operations.

3. *Addressing existing dams*

Opportunities exist to optimise benefits from many existing dams, address outstanding social issues and strengthen environmental mitigation and restoration measures. Dams and the context in which they operate are not seen as static over time. Benefits and impacts may be transformed by changes in water use priorities, physical and land use changes in the river basin, technological developments, and changes in public policy expressed in environment, safety, economic and technical regulations. Management and operation practices must adapt continuously to changing circumstances over the project's life and must address outstanding social issues.

4. *Sustaining rivers and livelihoods*

Rivers, watersheds and aquatic ecosystems are the biological engines of the planet. They are the basis for life and the livelihoods of local communities. Dams transform landscapes and create risks of irreversible impacts. Understanding, protecting and restoring ecosystems at river basin level is essential to foster equitable human development and the welfare of all species. Options assessment and decision making around river development prioritises the avoidance of impacts, followed by the minimisation and mitigation of harm to the health and integrity of the river system. Avoiding impacts through good site selection and project design is a priority. Releasing tailor-made environmental flows can help maintain downstream ecosystems and the communities that depend on them.

5. *Recognising entitlements and sharing benefits*

Joint negotiations with adversely affected people result in mutually agreed and legally enforceable mitigation and development provisions. These provisions recognise entitlements that improve livelihoods and quality of life, and affected people are beneficiaries of the project. Successful mitigation, resettlement and development are fundamental commitments and responsibilities of the State and the developer. They bear the onus to satisfy all affected people that moving from their current context and resources will improve their livelihoods. Accountability of responsible parties to agreed mitigation, resettlement and development provisions is ensured through legal means, such as contracts, and through accessible legal recourse at the national and international level.

6. *Ensuring compliance*

Ensuring public trust and confidence requires that the governments, developers, regulators and operators meet all commitments made for the planning, implementation and operation of dams. Compliance with applicable regulations, criteria and guidelines, and project-specific negotiated agreements is secured at all critical stages in project planning and implementation. A set of mutually reinforcing incentives and mechanisms is required for social, environmental and technical measures. These should involve an appropriate mix of regulatory and non-regulatory measures, incorporating incentives and sanctions. Regulatory and compliance frameworks use incentives and sanctions to ensure effectiveness where flexibility is needed to accommodate changing circumstances.

7. Sharing rivers for peace, development and security

Storage and diversion of water on transboundary rivers has been a source of considerable tension between countries and within countries. As specific interventions for diverting water, dams require constructive co-operation. Consequently, the use and management of resources increasingly becomes the subject of agreement between States to promote mutual self-interest for regional co-operation and peaceful collaboration. This leads to a shift in focus from the narrow approach of allocating a finite resource to the sharing of rivers and their associated benefits in which States are innovative in defining the scope of issues for discussion. External financing agencies support the principles of good faith negotiations between riparian States.

In order to achieve equitable and sustainable outcomes, free of the divisive conflicts of the past, future decisions on water and energy resource projects will need to reflect and integrate these strategic priorities and their associated policy principles in the planning and project cycles. The Commission considers that its recommendations can best be implemented by focusing on five key stages in decision making, which influence the final outcome and where compliance with regulatory requirements can be verified. A set of criteria is promoted for each decision point against which decision makers and stakeholder groups can check compliance with agreed procedures and commitments. The first two decision points relate to the planning phase:

– *needs assessment* - validating the needs for water and energy services;
– *selecting alternatives* - identifying the preferred development plan from among the full range of options.

Where a dam emerges from this process as a preferred development alternative, three further critical decision points occur:

– *project preparation* - verifying that agreements are in place before tender of the construction contract;
– *project implementation* - confirming compliance before commissioning;
– *project operation* - adapting to changing contexts.

The Commission presents 26 guidelines for practical implementation of the new direction for appropriate and sustainable development (Table 1). These guidelines are considered to be just that - guidance. They reflect good practice around the world that can be used in a tangible and practical way to meet the strategic priorities laid out by the Commission, and they can be used and adapted to different local circumstances. Additional guidance is available within the Commission's extensive knowledge base, which is available on its website: *www.dams.org* and on a CD-ROM.

The Commission's recommendations require more attention to be paid to the preliminary stages of planning. This has time and cost implications, but in the long run, the Report offers the opportunity to reduce conflict, reduce delays and lower overall costs to affected people, the government, the operator and to society in general.

Concluding remarks

The Commission did not set out to forecast future needs. Rather it concentrated on providing process recommendations through which individual countries decide their own priorities. The WCD report goes beyond principles to demonstrate how to ensure effective compliance, to help develop resettled communities and to protect ecosystems. The main challenge is to select appropriate options and to provide the means to achieve them - that is where the WCD has provided guidance. The following excerpt is taken from the 'Call to Action' in the final chapter of the Report.

'The work of the World Commission on Dams is over. But the concerns that led to its establishment are still with us. Dams have too often left a legacy of social injustice and environmental damage, and that legacy does not disappear because we have identified a better way of doing things in future. Early and resolute action to address some of the issues arising from the past will go a long way to building the trust required to enable the different actors to work together. So, too, would an assurance to countries still at an early stage of economic development that the dams option will not be foreclosed before they have had a chance to examine their water and energy development choices within the context of their own development process.

Table 1.   26 Guidelines for the planning and project cycle (World Commission on Dams, 2000).

Strategic priority 1: gaining public acceptance
1.   stakeholder analysis
2.   negotiated decision making processes
3.   free, prior and informed consent

Strategic priority 2: comprehensive options assessment
4.   strategic impact assessment for environmental, social, health, and cultural heritage issues
5.   project-level impact assessment for environmental, social, health, and cultural heritage issues
6.   multi-criteria analysis
7.   life cycle assessment
8.   greenhouse gas emissions
9.   distributional analysis of projects
10.   valuation of social and environmental impacts
11.   improving economic risk assessment

Strategic priority 3: addressing existing dams
12.   ensuring operating rules reflect social and environmental concerns
13.   improving reservoir operations

Strategic priority 4: sustaining rivers and livelihoods
14.   baseline ecosystem surveys
15.   environmental flow assessment
16.   maintaining productive fisheries

Strategic priority 5: recognising entitlements and sharing benefits
17.   baseline social conditions
18.   impoverishment risk analysis
19.   implementation of the mitigation, resettlement and development action plan
20.   project benefit-sharing mechanisms

Strategic priority 6: ensuring compliance
21.   compliance plans
22.   independent review panels for social and environmental matters
23.   performance bonds
24.   trust funds
25.   integrity pacts

Strategic priority 7: sharing rivers for peace, development, and security
26.   procedures for shared rivers

The experience of the Commission demonstrates that common ground can be found without compromising individual values or losing a sense of purpose. But it also demonstrates that all concerned parties must stay together if we are to resolve the issues surrounding water and energy resources development. It is a process with multiple heirs and no clear arbiter. We must move forward together or we will fail'.

While some groups have been critical of the viability, or acceptability of some of its recommendations, and some have said the report is 'anti-development' the Commission is clear that the decline in international finance and disputes over major projects will continue to make dams controversial until there is a sea change in the way in which dams are planned, designed and managed. It believes that only through mutual confidence building between the different parties will dams remain a legitimate and viable response to meet societies' needs and that development decisions should be taken with wider participation and accountability.

Despite interpretations that the report hands a 'veto power' over to the minority, this is not what the Commission advocates. The report recommends seeking the prior informed consent of indigenous peoples, with reference to judicial review in cases where no consent is possible. The supreme and independent role of the State as final arbitrator remains intact, despite complaints to the contrary. Nothing in the Report erodes this authority. Strengthening the active role of citizens and making government agencies more accountable is an integral part of democratic nation building

The WCD experience will reinforce a form of development that respects human rights, ensures equitable distribution of development benefits, and fits within a frame set by the need for environmental care. Far from being anti-development, Dams and Development is likely to be seen as reinforcing development through speeding the transition from traditional approaches based on political and financial power, to the new consensus emerging around the notion of sustainable human development.

Reference

World Commission on Dams, 2000, Dams and development: a new framework for decision making, *Earthscan*, London, Great Britain

6

ICOLD's criteria applied to the southern delta area

Storm surge barrier (Eastern Scheldt) and Dike Enforcement Project (Western Scheldt*)*

Ir. Leo Santbergen, Jan Willem Slager and Drs. Kees Storm

Introduction

In May 1997, the International Commission on Large Dams (ICOLD) published its position paper on dams and the environment (ICOLD, 1997). The organisation states that *'concern for the environment, including both natural conditions and social aspects, must be manifest from the first planning steps, throughout all phases of design and implementation, and during the entire operating life of a project'.* This paper presents two case studies in the Southern Delta area of the rivers Rhine, Meuse and Scheldt: the Eastern Scheldt storm surge barrier (completed in 1986) and the Dike Enforcement Project in the Western Scheldt (still in process). Figure 1 shows a map of the Delta area. The authors answer the question to what extend the planning and construction of the storm surge barrier and the dike enforcement meet ICOLD's criteria.

The first case: the Eastern Scheldt storm surge barrier

In 1976 the Dutch government decided to build a storm surge barrier in the mouth of the Eastern Scheldt estuary, instead of an 8 km long closure dam. The decision was primarily the result of a political compromise between flood defence and preserving the (marine) environment. By doing so, the Dutch created an enormous technical challenge and, to a certain extent, preserved national and internationally important nature values. Shellfish cultures of relatively large importance to industry in this region continued to prosper. Although the decision to build the barrier is considered a breakthrough on the way to integrated water management in The Netherlands, negative consequences on the marine environment do occur as well. Additional investments still have to be made aiming at the re-establishment of valuable estuarine gradients and at restoring tidal habitats. (Storm and Santbergen, 2000). Figure 2 shows an aerial photograph of the storm surge barrier.

If we look at ICOLD's criteria on dams and the environment successively, we come to the following review of this project.

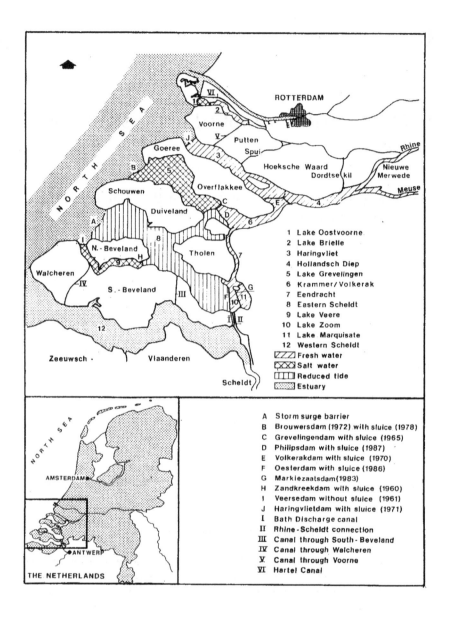

1 Lake Oostvoorne
2 Lake Brielle
3 Haringvliet
4 Hollandsch Diep
5 Lake Grevelingen
6 Krammer/Volkerak
7 Eendracht
8 Eastern Scheldt
9 Lake Veere
10 Lake Zoom
11 Lake Marquisate
12 Western Scheldt

▧ Fresh water
⬚ Salt water
⬚ Reduced tide
⬚ Estuary

A Storm surge barrier
B Brouwersdam (1972) with sluice (1978)
C Grevelingendam with sluice (1965)
D Philipsdam with sluice (1987)
E Volkerakdam with sluice (1970)
F Oesterdam with sluice (1986)
G Markiezaatsdam (1983)
H Zandkreekdam with sluice (1960)
I Veersedam without sluice (1961)
J Haringvlietdam with sluice (1971)
I Bath Discharge canal
II Rhine-Scheldt connection
III Canal through South-Beveland
IV Canal through Walcheren
V Canal through Voorne
VI Hartel Canal

Figure 1.   Map of the Delta area of the rivers Rhine, Meuse and Scheldt in the Southwest of The Netherlands.

Criterion a:

Concern for the environment, including both natural conditions and social aspects, must be manifest from the first planning steps, throughout all phases of design and implementation, and during the entire operating life of a project. During the initial stages of planning a dam project, the question should be studied whether alternative solutions exist that could possibly fulfil the various purposes of the dam project at lower long-term costs to society and the environment.

Figure 2.   Aerial photograph of the Eastern Scheldt storm surge barrier.

The so-called *Deltaplan* followed the disastrous flood in 1953. During the first planning steps of the Eastern Scheldt project in the early seventies, there was a strong political commitment to build dams in order to minimise flood risks in the shortest time possible. Therefore the idea was to build an 8 km long enclosure dam, creating a large freshwater lake profitable for agriculture. Alternative solutions didn't fit in this single-purpose management approach, focusing mainly on flood protection.

At the end of the roaring sixties, environmental awareness started to influence both interest groups and the government. Apart from the beneficial aspects of an enclosure dam, negative consequences were highlighted as well. The social-democratic government was sensitive to the public debate and appointed a special advisory committee in 1973. The committee studied alternative solutions. In 1976, after three years of complicated and often polarised discussions, the government decided to build a storm surge barrier (Knoester, et al., 1983).

Social aspects were taken into account during the entire decision making process, but the focus has changed form agriculture in the sixties to fisheries in the seventies.

During the eighties, researchers became more and more aware of the environmental and socio-economic consequences of the works. Many of these were not and could not have been foreseen in 1976 when the decision was made (Van Westen and

Vrijling, 2000). With present knowledge about costs and benefits an important question is if the storm surge barrier on the long term will turn out to be the best option for flood protection at the lowest costs to society and the environment.

Criterion b:

The enormous increase in human knowledge, including that in the field of environmental science, means that a whole team of specialists is needed to access and utilise that knowledge for a water resources development project.

In the original concept for a closure dam, planning and implementation was mainly a single-disciplinary, engineering approach. Since the early seventies, a significant evolution has taken place within the Ministry of Transport, Public Works and Water Management. An Environmental Division was established within the Delta Department of the Ministry in 1971. Since then, more and more multidisciplinary teams have emerged with ecologists, hydrologists, chemists, policy analysts, economists, etc.

Criterion c:

The larger the project, the greater the effects on the natural and social environment to be expected, and the wider the scope of the multidisciplinary, holistic study, which they require. Large-scale development demands integrated planning for an entire river basin before the implementation of the first individual project(s). Where river basins are part of more than one country, such planning presupposes international co-operation.

The Eastern Scheldt project has always been considered a large and difficult project, even the originally planned enclosure dam. For this reason, it was planned as one of the final closures, profiting from all the experiences with former closures. But as said before, the original scope was limited: safety first!

The Eastern Scheldt is part of the delta area of the rivers Rhine, Meuse and Scheldt. There is no integrated development and management plan for the entire catchment of these three rivers. The Deltaplan has caused disconnections among the upstream and downstream parts by the building of (inland) dams and barriers. As a consequence highly productive brackish water areas have been lost and barriers for migratory fish exist.

International co-operation takes place in the three Commissions on the Protection of the Rivers Rhine, Meuse and Scheldt. So far, these river basin authorities mainly focus on water quality improvement, ecological restoration and to a lesser extent on flood protection.

Criterion d1:

Projects must be judged everywhere and without exception by the state-of-the-art of the technologies involved and by current standards of environmental care. The scope for reducing any detrimental impacts on the environment through alternative solutions, project modifications in response to particular needs, or mitigating measures should be thoroughly investigated, evaluated and implemented. A comprehensive Environmental Impact Assessment, since 1971 mandatory in a growing number of ICOLD member countries, ought to become standard procedure everywhere as part of project conceptualisation, which is well before final design and the start of construction.

There was no official *Environmental Impact Assessment* in the case of the Eastern Scheldt project, but a comparable policy analysis process was followed. In 1976 a policy document was published with an analysis of mainly 3 alternatives:
- a complete closure;
- a closure with a storm surge barrier;
- no barrier but raising of height of the sea dikes.

These alternatives were rated for effects on: safety, environment, fisheries, water management, navigation, recreation, regional planning, costs, legal procedures and the effects on employment and economy (Ministry of Transport, Public Works and Water Management, 1976). The decision making process was influenced to a significant extent by great political tensions. Therefore, certain alternatives were studied only very briefly or speedily rejected. In retrospect, it seems at least surprising that the alternative to keep the Eastern Scheldt open and raising the height of the existing dikes was rated more undesirable whilst the adjacent Western Scheldt was kept open due to the important navigation channel to and from the harbour of Antwerp.

Criterion d2:

Countries still lacking in expertise on the legal framework and administrative structures should receive assistance from countries where the relevant legislation is more advanced and the necessary practical experience has been gained with regard to the extent of the investigations required, the methods and procedures to be employed, and the conclusions to be drawn from the results. Special attention should be paid to any effects on biodiversity or the habitat of rare or endangered species.

Before considering planning and implementation of projects similar to the Eastern Scheldt storm surge barrier in other countries with comparable water systems, problems better should be defined at the level of those entire water systems. *Possibly*, in some river basins, a construction like the storm surge barrier could be *one* of the alternative solutions.

Concerning biodiversity and the habitats of rare or endangered species, an extensive ecosystem-monitoring program has been carried out in the Eastern Scheldt. Elements of this program could possibly be integrated into monitoring programs in other countries.

Criterion e:

The decision on what is a very considerable investment for a dam project must be based on an unequivocally realistic economic analysis. Any tendency to overstate the benefits and understate the costs must be strictly avoided. This also requires taking the impact on the natural and social environment into account. Multipurpose benefits which do not produce revenues for financing the project must nevertheless be taken into account in the assessment of a project or a comparison with alternatives.

In our opinion, a cost-benefit analysis based on the economic and ecological values of a water system should actually be carried out for at least the life span of the construction, which is 200 years in the case of the storm surge barrier. It is vital that such an analysis is executed on an impartial basis. Political pressure can be very influential in the final results of the analyses. In the case of the Eastern Scheldt, we now observe some inconsistencies in the ratings of the alternatives. If we would have to establish a cost-benefit-analysis with present knowledge and methodology, the concerns of the 'no barrier' alternative would probably be estimated more positive, and the benefits of the 'storm surge barrier' more negative. Table 1 gives an example of a water-system-orientated cost-benefit approach for the storm surge barrier compared with the hypothetical alternative of a closure dam.

Table 1.   Example of a water-system-orientated cost-benefit approach for the open storm surge barrier compared with the (hypothetical) situation of a closed barrier.

Costs	Benefits
– higher construction costs	– preservation of the biodiversity of the tidal marine environment
– higher maintenance costs (flood prevention)	– preservation of nursery area for fishes
– longer construction period	– preservation of shell fisheries
– fewer optimal water recreation possibilities	– increased knowledge and experience in the fields of civil engineering and marine sciences
	– fewer eutrophication problems
	– greater possibilities for future water managers

One the major drawbacks of a cost-benefit approach is that we cannot predict the effects for such a long time scale and even if we could predict them, we don't know how future generations will appreciate these changes. Nonetheless, we believe that a water-system-orientated cost-benefits analysis can be an important tool in the planning and decision making process. It can be useful in clarifying the discussions, for instance, with interest groups.

Criteria f and g (resettlement):

Involuntary resettlement must be handled with special care, managerial skill and political concern based on comprehensive social research and sound planning for implementation. The associated costs must be included in the comparative economic analysis of alternative projects, but should be managed independently to ensure that the affected population will be properly compensated. For the population involved, resettlement must result in a clear improvement of their living standard, because the people directly affected by a project should always be the first to benefit instead of suffering for the benefit of others. Special care must be given to vulnerable ethnic groups.

Even if there is no resettlement problem, the impact of water resources development projects on local people can be considerable during both construction and operation. All such projects have to be planned, implemented and operated with the clear consent of the public concerned. Hence, the organisation of the overall decision making process, incorporating the technical design as a sub-process, should involve all relevant interest groups from the initial stage of project conceptualisation, even if existing legislation does not (yet) demand it. In the continuous, comprehensive and objective information transfer from planners to the public, dam engineers must contribute, through professional expertise, to a clear understanding and dispassionate discussion based on facts and not on irrational ideas of the positive and negative aspects of a project and its possible alternatives. Dam promoters must act as mediators and educators with the aim of becoming good neighbours and not intruders.

Over all, about 200 people had to resettle due to the Eastern Scheldt project. But marine cultures were also replaced. All companies or civilians who felt harmed by the project could apply for compensation. Even now, some legal procedures are still in the process of being settled.

In the original planning process of the Delta project, the information transfer from the Dutch government was a one-way communication in which safety against floods was emphasised strongly. Strong opposition from environmental organisations, fishermen and scientists forced the government into more openness towards the public. This openness initiated an evolution towards public involvement as a key-element in The Netherlands.

Criteria h and i:

A complete post-construction audit of an entire project or at least a performance analysis of major impacts should be carried out in order to determine the extent to which the environmental objectives of the project or of certain mitigating measures are being achieved. The results of such analyses should be published as a contribution to our knowledge on such matter, and for application to future projects.

As soon as a project becomes operational, its impact on the environment should be assessed at regular intervals, based on data and sources resulting from adequate pre-construction monitoring. Depending on the individual situation, certain critical parameters should be monitored as a basis for a subsequent performance analysis of the project, resulting in a better understanding of its interactions with the environment.

The extent to which the environmental objectives have been achieved was described by means of an ecological monitoring program, which was evaluated for the first time in 1991 and is repeated every 5 years (Iedema, et al., 1991).

A complete post-construction audit of the entire Delta project, including the storm surge barrier, has not been carried out. However, several aspects have been reviewed on the basis of present ideas of integrated water management.

Criterion j:

In this context, there is also a need for more ecological research on dams and reservoirs, which have already seen many years of service. Mistakes and shortcomings could be avoided. Many of the recurring controversies relating to the ecological impacts of new dam projects could be prevented. And the problems involved could be clarified and solved more easily if our latent store of long-term experience with the operation of so many dams and reservoirs were collected, processed, evaluated and published in the framework of research projects based on carefully directed investigations. Such research projects would also provide and enhance the basis for a general policy of intensified collaboration with environmental scientists.

It may be clear from this paper that we strongly support these final statements!

The second case: the Dike Enforcement Project (Western Scheldt)

The extreme floods of 1993 and 1995 caused an evacuation of more than 250,000 habitants in the eastern part of The Netherlands. These events lead to the political commitment to realise a dike strengthening work, which is called the Major Rivers Delta Plan.

At the end of 1996 and the beginning of 1997 there was no disastrous flood or great evacuation of people. Technical research showed out that about half of the dike revetments along the Dutch shoreline are estimated to be too light; they don't meet the safety standards. This alarming signal situation lead to the decision to start with the Dike Enforcement Project Western Scheldt.

If we look at ICOLD's criteria successively, we come to the following review for this project.

Criterion a)

In the Western Scheldt the safety problem was great. Therefore, directly after the stormy season in 1997, water-boards, provincial authorities and the Ministry of Transport, Public Works and Water Management closely co-operated to enforce the weakest 10 km. Within this first hectic year there was no time for extensive procedures or criteria as formulated by ICOLD.

In the first two years more emphasis was on social aspects than on environmental aspects. However, both aspects were manifest from the beginning. After deliberation with municipal authorities and environmental organisations an optimal compromise was found in a very short time.

The works will endure till 2015, including dikes in the adjacent Eastern Scheldt. Within the project alternative dike designs are studied seriously. Long-term alternatives like depoldering to create extra floodplains are not included in the Dike Enforcement Project for the Western Scheldt. Other projects like the Long Term Vision on the Development and Management of the Scheldt estuary (co-operation between Flanders and The Netherlands) do include such alternatives.

Social aspects include recreational use of the maintenance roads on the dike reaches. Areas which are very important as resting, breeding and feeding places for birds are and will be excluded from recreational use.

Criterion b)

Since 1997, multidisciplinary teams emerged with engineers, ecologists, water managers, jurists, communication experts, landscape architects, quality risk managers, etc. For example, new dike constructions are tested on environmental and technical aspects using outdoor experimental and existing dike reaches. Knowledge of bird-watchers is necessary because the Western Scheldt is a wetland of international importance for many different species. In and around the Dutch dikes a lot of culture historical aspects are present and assessed by landscape architects.

As an example, figure 3 shows a dike reach with a continuous vegetation gradient from the adjacent marsh to the top of the dike. Figure 4 shows an interrupted vegetation gradient due to the used stone material.

Figure 3.   Hellegatpolder (Western Scheldt, near Kloosterzande village): a dike reach with a continuous vegetation gradient from the adjacent marsh to the top of the dike.

Criterion c)

At this moment there is no multidisciplinary river basin authority that integrates technical, social and environmental aspects on the scale of the entire river basin. However, following a phased approach, the project constantly adepts itself to the circumstances as soon as new developments are determined and decided on.

The designs of the dike reaches are made for a period of 50 years. Global developments, as sea level rising, are taken into consideration. In the next project phases, transboundary activities as dredging and dumping for the maintenance of the navigation channel to and from the Scheldt harbours, plans for the development of controlled floodplains and the plan for the construction of a storm surge barrier near Antwerp must be taken into consideration too. In the same way the newest techniques and hydraulic models and arithmetic methods are used, based on the current safety approach. In the current situation the norm is an exceeding chance for a dike reach of 1 : 4,000. In the future the norm will be based on a so called 'dike ring area'; on the actual flooding changes and social risks per area. The project works in close co-operation with the Flemish region on the border crossing dike reaches. The dike improvement in The Netherlands has no adverse environmental affect for the Flemish part of the Scheldt estuary.

Figure 4.   Reigersbergsche polder (Western Scheldt, near Rilland village): an interrupted vegetation gradient due to the used stone material.

Criterion d1)

Due to the urgent need for dike enforcement the provincial authority decided not to ask for an Environmental Impact Assessment as long as environmental (and social) aspects are taken fully into consideration according to Dutch laws and European guide-lines.

After the first hectic year in which safety ruled the agenda, an Environmental Impact Inventory (1997) and a Landscape Vision (1998) followed. The Environmental Impact Inventory gives a survey of the environmental and social values and an assessment of various dike design alternatives. By the end of 1997 administrators decided that coming dike reaches to enforce at least would have to restore (and if possible even improve) the ecological values. Since then, one of the basic principles of the project is that all the dike reaches have to comply with the rules of the Environmental Impact Inventory.

Criterion d2)

The various investigations and monitoring programs supply specific knowledge on technical, social and environmental aspects. Possibly these programs and results can be integrated into research and monitoring programs in other countries for similar dike enforcement projects.

Criterion e)

The Dutch government made a rough economic analysis in 1996. During the last four years water-boards have surveyed the dikes in more detail. The result is a so-called 'detailed test', which gives a better insight in the problem. At the same time research programs are carried out. Agreement is reached how to divide the costs over the different public bodies. This summer a new economic analysis, based on a multi-criteria analysis will be available and presented to the Minister.

Criteria f) and g)

There is no resettlement problem. The sea defence works are carried out in order to prevent involuntary resettlement!
 All works are carried out and communicated with all parties concerned. Different kinds of communication activities will inform the involved citizens. Everyone can give his/her opinion in a public inquiry procedure.

Criterion h) and i)

The project includes several monitoring programs to follow the performance and development of the enforced dike reaches. Besides, experimental dike reaches with different constructions are monitored. Research emphases on vegetation development, presence of birds, the erosion processes of newly developed green dikes and so called 'clay-dikes'.

Criterion j)

It may be clear from this lecture that we strongly support the final statements of ICOLD's criterion j.

Conclusions

To build future dams and dikes that fit well in a sustainable development and management of river basins, we have to look back once in a while, systematically reviewing projects in the context of the time period and circumstances in which they were conceived, designed and built.
 Before the ICOLD criteria were published water management in The Netherlands developed itself from a single-purpose approach to a more integrated approach. In this context, the bold decision in 1976 to build a storm surge barrier instead of a closure dam can be interpreted as a turning point.
 Both cases show us that in theory, ICOLD's criteria are very useful to obtain a systematic approach in the planning, design and implementation of dams and dikes in which technical, environmental and social aspects are integrated. In practice, short-term political objectives influence the extent to which alternatives are seriously being studied.
 There is still a long way to go to obtain integrated planning for an entire river basin before the implementation of large-scale development projects like dams and dikes take place. International river basin authorities therefore should broaden their scope to formulate river basin development and management plans, as meant in the recent published European framework-directive on water management. The role and

management of existing and planned dams and dikes should be defined within the context of entire river basins in order to obtain a more sustainable situation in which projects are carried out at lower long-term costs to society and the environment.

Even though our knowledge in the fields of environmental and social sciences and civil engineering is extensive, we are not capable to foresee all future changes caused by our large-scale development projects. The challenge is to cope with these unexpected changes through a flexible and adaptive water management approach. We have to be modest in our expectations as to what degree we can control nature. In fact, to follow ICOLD's criteria, we should carry out extensive monitoring programs during the entire life span of a construction to evaluate long-term costs and benefits.

If more and more countries should follow the systematic approach of ICOLD's criteria, eventually sharpened by the conclusion of the World Commission on Dams, to our opinion our world would be much better of! Therefore, international professional organisations like ICOLD and ICID should stimulate the application of its criteria on dams and the environment more intensively, also within their own technical and annual meetings. More research and investments should be made on alternatives to large-scale development projects.

Finally, systematically following ICOLD's criteria on dams and the environment and the recommendations of the World Commission on Dams, loss of valuable water resources as well as future investments to restore economic and ecological functions of these resources can be (partially) prevented.

References

International Commission on Large Dams (ICOLD), 1997. *Position Paper on Dams and Environment.*

Iedema, W. ed., 1991. Safe tides. Evaluation of the Eastern Scheldt five years after completion of the storm surge barrier (in Dutch). Ministry of Transport, Public Works and Water Management, *report AX-91.089.* The Hague, The Netherlands.

Knoester, M., J. Visser, B.A. Bannink, C.J. Colijn, and W.P.A. Broeders, 1983. The Eastern Scheldt project. *Water Science Technology*, Volume 16, Rotterdam, p.p. 51 - 77.

Ministry of Transport, Public Works and Water Management, 1976. *Analyses of Eastern Scheldt alternatives* (in Dutch). The Hague, The Netherlands.

Storm, C. and L. Santbergen, 2000. A breakthrough in integrated water management in The Netherlands: the case of the Eastern Scheldt storm surge barrier. In: *20th Congress on Large Dams*, ICOLD, Beijing, China, Volume 3, Question 77, Paris, p.p. 715 - 732.

Westen, C.J. van and H. Vrijling, 2000. Evaluating the Dutch Delta project. In: *20th Congress on Large Dams,* ICOLD, Beijing, China, Volume 3, Question 77, Paris, p.p. 703 - 714.

7

Role of the consultants

DRS. A. LEUSINK

Introduction

In The Netherlands systematic intervention in natural water conditions started already ten centuries ago. Through the building of mounds, dams and dikes the inhabitants of the lowlands enclosed tidal embayment and reclaimed land in the tidal areas and estuaries. At present without the protection of dunes, dikes and dams about 65% of the country would be inundated either by sea at storm surges or by the rivers at high discharges. Flood disasters caused by storm surges and severe river floods occurred in The Netherlands several times each century, even during the second half of the 20th century. The occurrence of such destructive events is mostly the decisive factor that forces government and responsible authorities to take action and to improve protection measures against floods.

The first structural water works in The Netherlands were applied at the reclamation of peat areas. The surface area in those cultivated lands, which were drained by ditches, entered into a process of irreversible land subsidence. They were turned into polder areas to provide those low-lying peat areas from flooding. In later years part of the vulnerable reclaimed land had to be given back to the sea. The inhabitants tried to protect those areas against floods through the building of dikes but they were not always successful. Besides the struggle against the sea there was a permanent threat for floods from the main rivers. However the rivers and estuaries also were most important waterways for transport of goods from the seaports to the hinterland.

Since the systematic intervention in the natural water situation the responsibility for the local and regional water management was entrusted to water-boards. At the end of the 18th century more than thousand authorities were engaged with water management in The Netherlands. The disintegration of authority caused serious problems for the dewatering of the area and the management of the sea defence works against floods. Therefore in 1798 the government of the first totalitarian State of The Netherlands created a centralized authority for the management of water. Engineers of this body became very influential with reference to the water management on central and local level.

The centralized authority became the actual Directorate-General of Public Works and Water Management (Rijkswaterstaat) which is not only responsible for the management of the water works but since the 20th century they also focus a lot of attention on scientific and applied research. Teams of state civil engineers planned and designed the water works assisted by experts of technological institutes. Consequently until the 1970's the involvement of the private consulting engineers in water works in The Netherlands has been relatively low, although they had gained broad experience in hydraulic engineering abroad. The situation changed due to the introduction of an integrated approach in water management and the consciousness for our natural environment. Since the 1980's consultants play an important role in many ongoing works related to water and transport infrastructure, not only on technical but also on socio-economic and environmental aspects.

Position and network

Consulting engineers want to provide their services to better structure and manage the natural and built-up physical environment. Their advice is based on extensive multi-disciplinary knowledge, skills, experience and information from similar problems elsewhere. They play an important role in the transfer of knowledge from universities and research institutes to the application in practice. Consulting engineers apply it and continuously enrich the knowledge for the solution of concrete problems. In reverse they also translate daily problems with the planning, design and management of infrastructure into specific questions for specialized institutes and universities.

In The Netherlands there are more than 250 consulting engineering firms with a total staff of 25,000. They encompass a very broad range of disciplines and are active for all stakeholders in the water sector: government, private sector, international agencies, NGO's, utility sector, technological institutes and the citizens. Their competencies are multiple and consequently their influence has increased enormously in the recent past. They apply a holistic integrated approach and know how to solve complex and multi-disciplinary problems. Due to their actual role at the home market referred to water and environment Netherlands firms could become in the last decade one of the top five international consulting engineering communities in the world.

Role in the project cycle

Consultants play an important role in many water projects all over the world. They deliver their expertise as project manager or as an expert in a specific field of interest. In the past their role was mostly limited to the planning and design phases of projects. In the last decades however they became involved in all stages of the project cycle, from policy preparation to management of the works. Clients, which make use of their services, profit from their lessons learned in other projects and their knowledge of the latest technology. Consulting engineers' role is most prominent in the following aspects:
- *policy formulation*, basic principles are described and the role of dikes and dams is illustrated; demand management studies and small-scale alternatives are included;
- *idea development*, this means in many cases the very beginning of a project. Client's problem statement is clearly formulated and conceptual designs of civil works are presented;
- *decision making*, consultants provide objective and rationale arguments and deliver pre-feasibility level results or run the models for scenario comparison.

All aspects are included: technical, economical, social, environmental, financial, etc. Public acceptance of the proposed plan is crucial and will be guided;
- *planning phase,* execution of feasibility studies and cost-benefit analyses;
- *design of works;*
- *supervision of works;*
- *risks assessment.*

Basic information needed to give follow-up to the recommendations made by the World Commission on Dams (WCD) is collected by consultants and reported to their clients. The Terms of Reference of the client determine to a large extent the contents of the studies. However a code of conduct of the consultancy firm encourages the individual advisors to present the outcome of the studies in the most objective way. The global debate on dams relate to the functioning of dams and the influence of the works on the natural river system and consequently the affected people: Will the river flow or the rights of access to the water resources be distorted, or the environmental resources be degraded, or the living conditions of the local communities be threatened, and is this the best economic investment of public funds. Consultants in The Netherlands are aware of the fundamental implications of sustainable development options for the execution of their professional activities. Today's solutions to fulfil the basic needs should not limit future generations to make their living. In the framework of the International Federation of Consulting Engineers (FIDIC), their intentions have been elaborated in guidelines, manuals, codes of conduct, etc, that are imitated world wide.

Involvement of the consultant

Consultants have to comply with their contracts and the consultant cannot do much with respect to transfer of technology unless this is part of the contract. They have to convince their clients that sustainability and transfer of knowledge is an integral part of the job. They should develop a professional attitude towards the transfer aspect as well as towards other objectives of the contract, particularly by giving due responsibility to the local partner. For effective transfer of technology, the foreign consultants' aim should be to guide the local consultants to do the work, rather than actually doing it. Local consultants depend on the contract just as their international colleagues do. They also have a role and responsibility in the creation of a reliable consultancy sector in their countries. There may be a need to increase awareness of the craft and culture of engineering, with emphasis on their societal role. To be effective in the long-term, however, it is important that transfer is undertaken on a fully integrated team or firm basis rather than simply to an individual who may not have a full range of local support.

There are many ways to achieve transfer of knowledge and each method has a particular advantage or disadvantage depending on many factors. It is very important to carefully select the appropriate method in view of the local circumstances and objectives. Notwithstanding the method of approach, an effective transfer of knowledge through projects will hardly be possible unless the receiving party plays a meaningful role in the project. Full integration of local and foreign experts in one project team will ensure the local input necessary to make transfer of knowledge work. The transfer of know how should be aimed not only at passing on technical skills, but also on the overall aspects of project development and management, environmental concern and socio-economic consequences.

Dikes and dams projects in many cases deal with transboundary aspects, which require co-operation between riparian states. Many conventions have been drawn-up and ratified to facilitate common action or to set criteria for the development of plans for the improvement of water management in one country. This is especially the case in river projects with upstream-downstream positions, both with respect to the availability of water and floods, as well as the pollution of the water. Upstream-downstream thinking of governmental agencies is often synonymous with acting on strong and weak positions. Mixed consultants project teams may come forward with other approaches, such as the principle of 'shared resources'. It is a prerequisite that both parties integrate their data and perceive that they deal with consistent and reliable data. This may lead to a better understanding of each other's position and an integrated water resources management approach at the level of the whole river basin.

Interventions in river basins or estuaries provoke lengthy debates about the valuation of nature. Sustainable economic development urges decision makers to make trade-offs between economic and ecological interests. Up to now there is poor experience with instruments for the valuation of nature in river basins and coastal areas. In recent years Decision Support Systems have been developed and case studies have been executed on a pilot scale, but broad experience with the application in projects is lacking. In this very important field of interest the role of the consultants may be of great value. Cost-benefit analyses and monetary valuation of nature require large data sets, which may be collected from prior consultants' studies for other purposes. Also practical experience with the application of such tools is necessary in case of use elsewhere, because it is impossible to walk around with blueprints. You always have to anticipate different ecological and socio-economic values.

Netherlands expertise abroad

Under the umbrella of the coordinating body NEDECO, since the 1950's Netherlands consulting firms cooperate to execute large projects abroad. In this framework also the specific expertise of governmental institutes and water-boards is made available for overseas clients. Experience gained with the execution of the big Lake IJssel and Delta projects appears to be of great value for their involvement in water development projects in the downstream part of river basins or in deltaic areas in Korea, Vietnam, Thailand, Indonesia, Bangladesh and others. Large-scale subsurface drainage and irrigation projects have been realized in Egypt, Iraq, Pakistan, Indonesia and India.

Over the last five years Netherlands consultants have been actively involved in the development of plans for the construction of a closure dam at the Gulf of Khambhat. The fresh water stored in the lake is to be used to supply domestic and industrial water to the drought-prone areas of Gujarat. The project also includes large-scale irrigation development, land reclamation and development of the potential for tidal power. The economic feasibility of the project increases enormously through the exploitation of the large tidal range of 8 to 11 m in the gulf to generate power. The closure dam designs show a length of 30 km across the Gulf of Khambhat and about 40 km across the Narmada estuary. Using conventional methods will not suffice to close off the gulf. Tidal currents could be as high as 10 m/s if a straightforward horizontal or vertical closure would be attempted. Therefore a technical feasible solution proposed, which have been applied already successfully

in The Netherlands, would be a combination of gradual horizontal and sudden vertical closure by caissons. This ambitious project envisages providing:
–   drinking water for over 15 million people at the peninsula of Saurashtra who currently face prolonged water shortages;
–   irrigation water for over a million hectares of arable land;
–   reclamation of some 120,000 hectares of currently saline tidal flats;
–   tidal power, depending on the installed capacity a maximum of 12,000 GWh of reliable and renewable energy could be provided annually;
–   improved transport facilities by road and water.

Another important ongoing dam project concerns the completion of the storm-surge barrier for the city of St. Petersburg in Russia. NEDECO in co-operation with the Netherlands ministry for Transport, Public Works and Water Management plays an active role in the relaunch of construction works of the 25 km long barrier with two navigation channels and six sluices. The dam and the sluices are completed for eighty per cent but the closures of the navigation channels still are in a very early stage. The annual flooding of the city centre, which causes severe damage to the properties, will come to an end after the completion of the works. Environmental effects in the Gulf of Neva have been studied thoroughly and negative side effects may be avoided through normal operation practices.

References

FIDIC, 1999. Expanding the boundaries, Speakers notes of the annual meeting held in The Hague 19 to 24 September 1999.
Huisman, P., et al., 1998. *Water in The Netherlands*, 2nd edition.
Ruijgrok, E.C.M., 2000. Valuation of nature in coastal zones. PhD Thesis.

8

Contractors, their changing role

Ir. J.J. Enneking

The historic role of the contractor in dam building, and also in other large hydraulic construction projects, is to sit and wait for the tender documents to appear, to prepare his best bid and to build the project in accordance with the contractual documents, thereby trying to earn some money in the process.

If we look at the (5) Key Decisions Points as defined by the World Commission on Dams (WCD):
1. Needs Assessment;
2. Selecting Alternatives;
3. Project Preparation;
4. Project Implementation;
5. Project Operation;

this 'historic' role of the contractor is found only in Point no. 4: Project Implementation.

However times have changed:
- the 'easy' project-sites were used in the 60's and 70's, many projects are now characterised by, for instance, difficult foundation conditions, less favourable access and exposure to more severe climatic conditions;
- world-wide accepted environmental rules and all kind of QA-norms (ISO-type) have come to life and responsible contractors are committed to observing them;
- contractors have developed and have invested in all kind of high-tech and large capacity construction equipment, which may be sometimes even project specific;
- trade-off between construction cost and maintenance cost is better understood, and also regarded as an important factor of the success of a project.

We also have to keep in mind that the present-day contractor:
- has developed substantial design capability;
- has direct, practical experience with observing environmental regulations, particularly during construction;
- has an interest in reducing risks, or putting the risks where they logically belong;
- may have access to tied export-credits.

These 'changed' times have lead to contractual forms which are sometimes very different from the old FIDIC division of roles between client, consultant (= designer and supervises construction), contractor ('constructs' what is in the documents).

Without being complete we can list:
– classical tender, but allowing the contractor to propose (technical) alternatives and when cheaper allowing them in to the selection process;
– the project EIA-approval is the contractors responsibility, not the clients;
– the contractor is asked to provide guarantees for the 'success' of the project, such as water quality guarantees in clean-up projects, and long term maintenance obligations;
– design and construct;
– turnkey projects;
– risk-sharing arrangements, such as for instance alliance contracts;
– build-operate-transfer projects;
– concession contracts;
– all kind of other combinations of construction and financing of projects.

Out of hundreds of possible examples, three current contracts have been selected and described shortly hereunder, where pragmatic solutions were chosen, which in our opinion addressed the sharing of risks in a creative and correct way:
1. *Bangladesh, Gorai River diversion*
   The Gorai diverts water from the Ganges to the south during the high water season, but falls dry in the low water period. By dredging the bar at the Ganges-Gorai entrance and stimulating the river-scour, the Contractor successfully managed (for the 3rd year already) to keep the Gorai open. When asked to 'guarantee' the dry-season Gorai-flows, he refused: there are risks beyond the contractor's control.
2. *Egypt, Port Said New Port, design-and construct*
   The project was tendered upon a (basic) reference-design. The applicants were supposed to check/amend/optimise this basic design and make a lump sum offer. Final design was made after contract award. Originally maintenance was included in the tender package, but this was taken out in a later stage.
3. *Netherlands, Betuwelijn Gorinchem-Sliedrecht*
   This contract for the substructure of a high-speed rail link contains plenty of risks, which are not directly within the contractor's control (existing cables and pipelines, right of way problems, foundation conditions, etc). In order to come to a reasonable sharing of the project-risks, client and contractor placed part of the contract sum in a special fund; this fund (provision) is managed by a board in which both client and contractor are represented. The contract is called an 'alliance-contract'.

Let us have a look at where in the WCD Key Decision Points in the project cycle; the 'modern' contractor can have an added value compared to the classical role:
   1. selecting Alternatives (costing of various options to arrive at the cheapest overall solution);
   2. project Preparation (lowest cost to reach the desired result including use of very specific work methods and equipment);
   3. project Implementation (as before);

4. project Operation (low operation and maintenance costs).

How this input can be best used in real life, for each individual project, is a matter that will evolve in the next few years.

These are the advantages of more contractors input:
– better insight in project cost in an early stage, less cost overruns and/or surprises, and quite possibly lower project cost in the end;
– less risk of 'environmental' problems particularly during construction;
– shorter project preparation and construction times.

There are clearly also disadvantages:
– the involvement of contractors in early stages may be contractually 'difficult' to manage;
– a contractor may refuse to make his design capabilities available before tender;
– tied financing versus tenders for all comers;
– evaluation of tender results may become more difficult and open to manipulation;
– potential high up-front costs to unsuccessful contractors may occur.

Having studied the WCD-report and considering the above, we think that the following conclusions can be drawn for now:
– there are obvious advantages and disadvantages of a greater involvement of contractors in developing a project; the advantages outweigh the disadvantages;
– in general it is our opinion that the lines as set out in the WCD report, appear to be a step back on a road that today's construction industry has already entered.

List of contributors

Ir. Hans van Duivendijk is chairman Netherlands Committee on Large Dams.

Prof. Kaare Hoege was President of the International Commission on Large Dams (ICOLD) from May 1997 until September 2000. Under his presidency ICOLD has shown more and more interest in other than purely technical aspects related to dams, like environment, the river basin as a point of departure for study of a dam project and the policies required for the development of shared transboundary river basins.

At present **Prof. Dr. Bart Schultz** is the President of the International Commission on Irrigation and Drainage (ICID) with its Central Office in Delhi, India. Apart from that he is Head of the Environmental Engineering Department in the Directorate-General of Public Works and Water Management, Utrecht, The Netherlands and also Professor in Land and Water Development, International Institute for Infrastructural, Hydraulic and Environmental Engineering (IHE), Delft, The Netherlands. Prof. Schultz graduated from the Delft University of Technology, Delft, The Netherlands. He also obtained his PhD degree here on the topic 'Water Management of the drained lakes in The Netherlands'. His career includes more than 25 years of research, advising and project implementation in the field of land and water development, drainage, irrigation and environmental engineering. His extensive know how and experience is, may be, best be illustrated by mentioning that he is author of more than 140 articles in the field of land reclamation, drainage and irrigation and that he, during his career, visited more than 30 countries to appraise, evaluate, or advice on land reclamation, drainage and irrigation projects.

Henk Saeijs was probably the first one in the World who integrated technical, ecological and policy aspects of large hydraulic engineering projects by introducing the integrated water system approach. He was very much involved in the studies and mitigating measures in the environmental field for the Dutch Delta Project. In fact his PhD Thesis in 1982 was called 'changing Estuaries'. In ICOLD he has been one of the driving forces behind the basin-wide approach for dam siting, sizing and operation. Apart from his work as Chief Engineer - Director for the Public Works Department in the South - Western Delta of The Netherlands, he is also, since 1994, Professor Water Quality Policies and Sustainability at the Erasmus University in Rotterdam.

Kirsten Schuijt works at the Erasmus University in Rotterdam, The Netherlands. She has a Master's degree in Economics and is currently doing a PhD in Environmental Science. Her research focuses on the process of economic valuation of ecosystems in water management

Prof. Han Vrijling started his career by working for a contractor but very soon afterwards he got involved in the design of the Eastern Scheldt Storm Surge Barrier. There, he developed the probabilistic approach to design features for this huge hydraulic engineering project. As a member of the project management team he became finally responsible for the extensive research for this project. In 1989 he became professor at Delft University in the Faculty of Civil Engineering, initially only to lecture in probabilistic methods to be used in hydraulic engineering but at present he is one of the few full time professors in Hydraulic Engineering at Delft University of Technology.

During the past two years **Mr. Jamie Skinner** was a member of the Programme Staff at the Secretariat of the World Commission on Dams in Cape Town, South Africa. During the past months he has been active in disseminating the Report 'Dams and Development' prepared by the Commission. This Report was launched on the 16th November of last year in London. As the Secretariat of the WCD is due to close on 31st March we are probably only just in time to have him present here.

Mr. Leo Santbergen is, like Henk Saeijs, an example of the multi-disciplinary approach we Dutch have developed for designing and operating hydraulic engineering projects and water systems. Mr. Santbergen studied biology at the Agricultural University Wageningen and is, since 1992, working in the south western Delta for the Directorate-General of Public Works and Water Management in the Division for Integrated Water Management. He is also policy adviser to the International Commission for the Protection of the river Scheldt, with focus on water quality and ecology.

Jan Willem Slager, Directorate-General of Public Works and Water Management. Directorate Sealand. Dike Enforcement project Western Scheldt.

Kees Storm. Directorate-General of Public Works and Water Management. Directorate Sealand. Co-ordinator water management Eastern Scheldt.

Mr. Aalt Leusink studied Earth Sciences at the University of Amsterdam. Between 1972 and 1981 he was a researcher and lecturer at the Institute of Earth Sciences, at the Vrije Universiteit, Amsterdam, in the fields of hydrology, meteorology and computer sciences. Between 1981 and 1984 he was a Consulting Engineer with IWACO (Consultants for Water and Environment), and from 1984 - 1996 a Member of the Board of Directors. In 1996 Mr. Leusink became Managing Director of Schiphol Management Services (SMS), Amsterdam, and since 1997 he has been Managing Director of NEDECO, the umbrella organisation of Dutch consultants.

Mr. Jan Enneking works at present for the large Dutch dredging contractor Boskalis International as manager of the Marketing and Project Development Department and, having heard that, you may ask yourself what is he doing here in this world of dams and dikes. Well, that is simple: Mr. Enneking, some 25 years ago, was working for the Dutch Consulting firm HASKONING and there he was very much involved in the multi purpose aspects of a large dikes and dams project in the Sebou basin in Morocco. In fact the reason that a certain large dam not only was built in the end but also considerably increased in size was mainly due to his clever integration of technical and economic aspects and the multi purpose approach advocated.

Committees

Netherlands Committee on Large Dams (NETHCOLD)

Ir. Hans van Duivendijk, chairman
Prof. drs. Ir. J.K. Vrijling, Delft University of Technology, vice-chairman
Ir. C.J. van Westen, Directorate-General of Public Works and Water Management, secretary
Ir. J. Timmerman, Royal Haskoning, treasurer
Drs. A. Leusink, NEDECO
Ir. R.J. de Jong, Delft Hydraulics
Ir. P.W. Besselink, DHV Environment and Infrastructure
Ir. C. Stigter
Prof. B. Petry, International Institute for Infrastructural, Hydraulic and Environmental Engineering (IHE)

Netherlands National ICID Committee (NETHCID)

Ir. J.H. van der Vliet, Principal water-board Amstel, Gooi and Vecht, chairman
Ir. J.L. Terwey, secretary
Ir. F. Folkertsma, Association of Water-boards, treasurer
Ir. K.J. Breur, Delft University of Technology
Ir. F. Rutgers, Directorate-General of Public Works and Water Management
Prof. dr. ir. E. Schultz, Netherlands representative ICID

Netherlands Executive ICID Committee

Prof. dr. ir. E. Schultz, Directorate-General of Public Works and Water Management and International Institute for Infrastructural, Hydraulic and Environmental Engineering (IHE)
Ir. G. Uittenboogaard, DHV Agriculture and Natural Resources BV
Ir. C.W.J. Roest, Alterra
Prof. ir. R. Brouwer, Delft University of Technology
Ir. I.A. Risseeuw, Arcadis-Euroconsult
Ir. T. van der Zee, Government Service for Land and Water Use
Ir. P.J.M. Kerssens, Delft Hydraulics
Ir. J.R. Moll, Royal Haskoning
Dr. W.F. Vlotman, Alterra-ILRI
Ir. G.H. van Vuren, Wageningen University

T - #0287 - 101024 - C0 - 246/174/5 [7] - CB - 9789058095411 - Gloss Lamination